The Hand of the Engraver

SUNY SERIES, INTERSECTIONS: PHILOSOPHY AND CRITICAL THEORY

RODOLPHE GASCHÉ, EDITOR

THE HAND OF THE ENGRAVER

Albert Flocon Meets Gaston Bachelard

Hans-Jörg Rheinberger

Translated by Kate Sturge

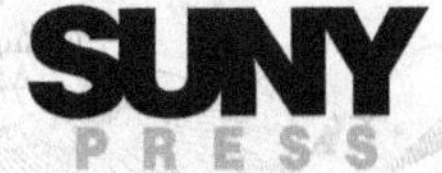

SUNY PRESS

Cover image, plate 16, "Le rideau" (28 x 19.5 cm), in Bachelard and Flocon, *Châteaux en Espagne*, 59. Courtesy of Catherine Ballestero.

Published by State University of New York Press, Albany

For information, contact State University of New York Press, Albany, NY
www.sunypress.edu

Library of Congress Cataloging-in-Publication Data

Names: Rheinberger, Hans-Jörg, author. | Sturge, Kate, translator.

Title: The hand of the engraver : Albert Flocon meets Gaston Bachelard / by Hans-Jörg Rheinberger ; translated by Kate Sturge. Other titles: Kupferstecher und der Philosoph. English

Description: Albany : State University of New York Press, 2018. | Series: SUNY series, intersections: philosophy and critical theory | Includes bibliographical references and index.

Identifiers: LCCN 2017059915| ISBN 9781438472119 (hardcover : alk. paper) | ISBN 9781438472126 (e-book) | ISBN 9781438472102 (paperback : alk. paper)

Subjects: LCSH: Flocon, Albert—Criticism and interpretation. | Bachelard, Gaston, 1884-1962. | Art and philosophy—France—History—20th century.Classification: LCC NE654.F59 R4913 2018 | DDC 769.92 —dc23

LC record available at https://lccn.loc.gov/2017059915

10 9 8 7 6 5 4 3 2 1

CONTENTS

List of Illustrations ⇴ vii

Acknowledgments ⇴ xi

Introduction ⇴ xiii

Surrationalism ⇴ 1

Albert Flocon, Gaston Bachelard ⇴ 5

Perspectives ⇴ 9

Reveries of Earth ⇴ 15

In Praise of Hands ⇴ 17

Landscapes ⇴ 23

On Engraving ⇴ 43

Poetics ⇴ 51

Castles in Spain ⇴ 57

Science, Art, Literature ⇴ 77

Notes ⇴ 83

Bibliography ⇴ 101

Index ⇴ 109

ILLUSTRATIONS

Figure 1a ⇴ 10

Plate 1 in Paul Éluard and Albert Flocon, *Perspectives: Poèmes sur des gravures de Albert Flocon*

Figure 1b ⇴ 11

Proof of 1a, single page

Figure 2 ⇴ 14

Plate 10 in Éluard and Flocon, *Perspectives*

Figure 3 ⇴ 19

Plate 7 in Gaston Bachelard, Paul Éluard, Jean Lescure, Henri Mondor, Francis Ponge, René de Solier, Tristan Tzara, and Paul Valéry; engravings by Christine Boumeester, Roger Chastel, Pierre Courtin, Sylvain Durand, Jean Fautrier, Marcel Fiorini, Albert Flocon, Henri Goetz, Léon Prébandier, Germaine Richier, Jean Signovert, Raoul Ubac, Roger Vieillard, Jacques Villon, Gérard Vulliamy, and Albert-Edgar Yersin, *À la gloire de la main*

Figure 4a ⇴ 26

Frontispiece in Gaston Bachelard and Albert Flocon, *Paysages*

Figure 4b ⇴ 28

Albert Flocon, preliminary drawing for the *Paysages* frontispiece

Figure 4c ⇴ 29

Albert Flocon, drawing for the *Paysages* frontispiece

Figure 4d ⇴ 30

Albert Flocon, frontispiece in the deluxe series of *Paysages* on China paper

Figure 5a ⇴ 32

Plate 4 in Bachelard and Flocon, *Paysages*

Figure 5b ⇴ 34

Drawing inserted into copy no. 12 of *Paysages*

Figure 6 ⇴ 35

Notebook page with pencil script, the reverse of the page of sketches reproduced in figure 5b

Figure 7a ⇴ 37

Plate 5 in Bachelard and Flocon, *Paysages*

Figure 7b ⇴ 38

Albert Flocon, preliminary sketch for plate 5

Figure 7c ⇴ 39

Albert Flocon, preliminary sketch for plate 5

Figure 7d ⇴ 40

Deluxe series on China paper, plate 7, in Bachelard and Flocon, *Paysages*, inserted into copy no. 7, detail

Figure 8 ⇴ 45

Plate “Le trièdre” in Albert Flocon, *Traité du burin. Avec une préface de Gaston Bachelard*

Figure 9 ⇴ 46

Plate “La charrue” in Flocon, *Traité du burin*

Figure 10 ⇴ 48

Plate “La caresse” in Flocon, *Traité du burin*

Figure 11 ↬ 58

"Grolierii Amicorum," printed invitation to a banquet on February 15, 1958, detail of frontispiece

Figure 12a ↬ 61

"L'auteur," plate 2, frontispiece in Gaston Bachelard and Albert Flocon, *Châteaux en Espagne*

Figure 12b ↬ 62

Preliminary study for the portrait of Gaston Bachelard

Figure 12c ↬ 63

Preliminary study for the portrait of Gaston Bachelard

Figure 13a ↬ 65

Sketch entitled "faire 10 châteaux en Espagne"

Figure 13b ↬ 66

Table of plates in Bachelard and Flocon, *Châteaux en Espagne*

Figure 14 ↬ 67

Plate 5, "Le nid d'aigle," in Bachelard and Flocon, *Châteaux en Espagne*

Figure 15a ↬ 71

Plate 16, "Le rideau," in Bachelard and Flocon, *Châteaux en Espagne*

Figure 15b ↬ 73

Ink sketch for "Le rideau"

Figure 15c ↬ 74

Ink drawing of "Le rideau"

Figure 16 ↬ 75

Bachelard and Flocon, *Châteaux en Espagne*, presentation of the drawings, ink sketch

ACKNOWLEDGMENTS

I would like to thank first of all Catherine Ballestero, Albert Flocon's daughter, who generously offered me access to letters and sketches from her father's papers. A very helpful source for research on Flocon's written oeuvre is her annotated catalog of his books and other writings: Catherine Ballestero, *Albert Flocon dans ses livres* (Neuchâtel: Éditions Ides et Calendes, 1997). I am grateful to Peter Schöttler for pointing me to documents in Mme Ballestero's possession. According to a communication from Mme Ballestero, Flocon's papers are now all held by the Institute for Contemporary Publishing Archives (IMEC) in Normandy. Gaston Bachelard's papers, as far as I could discern, are still not accessible even half a century after his death. I thank the former librarian of the Max Planck Institute for the History of Science, Urs Schöpflin, for kindly allowing me to use a drawing by Flocon from the copy of *Paysages* held by the institute's library. My thanks also go to Michael Hagner, Henning Schmidgen, Ineke Phaf, and Peter Geimer for attentive readings of earlier versions of this text. And last but not least, thanks to Kate Sturge for the translation.

INTRODUCTION

De lui à moi, pas de discours.
—*Gaston Bachelard*

"1933: the thunderclap of the thousand-year Reich, that intolerable prospect. . . . So into exile, with wife and child, destination Paris!"[1] Writing three years before his death in 1994, Albert Flocon is describing how his younger self, Albert Mentzel, experienced his exodus. He would spend the rest of his life in France.

This essay is about the encounter between Albert Flocon (1909–1994) and Gaston Bachelard (1884–1962), which began in the art circles of postwar Paris and continued for somewhat more than a decade, from the late 1940s to the end of the 1950s. The episode and the works to which it gave rise have attracted no detailed attention either from the philosophy of science or from the history of art.[2] They form the object of the present essay. A number of years ago, starting from an interest in Bachelard's epistemology, in particular his views on experimentation, I became aware of this connection between an engraver and a philosopher of science, a theorist of perspective and a theorist of poetics. A closer look at their works soon convinced me that this was a unique opportunity to investigate the interplay of hand and matter in poetic writing, in the art of engraving, and in scientific experimentation; it would allow me to explore the links and contiguities between those activities. My longstanding interest in the history and epistemology of the experiment as a form of rationality that is essentially embodied—in the hand of the experimenter as well as in the objects of manipulation—would find, so I hoped, new food for thought in this peculiar encounter. And I realized it would enable me to bring the two sides of Bachelard's oeuvre, his studies of the contemporary sciences and his reflections on literary writing, to bear on each other. Usually, Bachelard's epistemology and his poetology are kept in different boxes. Here, they appeared to me to come together on the terrain of an engraver's art, in the reflections of its making and the stories to which it gave rise.

The short chapters of this essay follow the encounter between Bachelard and Flocon in an essentially chronological manner, and they are organized themselves as encounters, each with one of the works initiating or resulting from their collaboration. The chapters provide glimpses into each of these works—that is, they proceed in an exemplary rather than an exhaustive fashion. My study can also be read as a contribution to an archaeology of situated knowledge in general, and of tactile knowledge in particular. Hands figure prominently throughout Flocon's work and Bachelard's musings on the elements, and so do they here. Above all, however, the piece is an homage to the two protagonists.

SURRATIONALISM

In 1936, a short programmatic essay on "surrationalism" appeared in the journal *Inquisitions*, edited by Louis Aragon, Roger Caillois, Jules-Marcel Monnerot, and Tristan Tzara. Its author, Gaston Bachelard, chose the following chiastic sentence to describe the relationship between the arts and sciences of his day: "An *experimental reason* will be established, capable of organizing reality surrationally as the *experimental dream* of Tristan Tzara organizes poetic liberty surrealistically."[1]

Bachelard is here alluding to Tristan Tzara's extended essay *Grains et issues* (Seeds and bran), published a year earlier. Its first chapter, "Rêve expérimental," addresses the phenomenon of *écriture automatique* and the relationship between daydreams and focused thought.[2] In his lengthy note on the chapter, Tzara explains what he means by an experimental dream, which has to do with the production of poetic texts and thus the representation of a structure that makes it possible to "bring forth new events not foreseen by the original plan."[3] Tzara's note refers to the birth and progress of his own text, which he recounts as follows:

> Thus the story follows, and spreads across the frame of a logical development that reduces itself to an account of successive facts, leaving an irrational and lyrical remnant open for discovery. This, in turn, overflows the vessel intended for it, and at times engulfs and floods the base, the foundation, the traditional scaffolding of the story. It is

> a lyrical superstructure whose elements are derived from the base structure and which, once it is realized, impacts back onto that structure from the heights of its new power. Occasionally, its force intensifies to such an extent that it undermines the meaning of the structure, corrupts it, abolishes it, annihilates it in its essence.[4]

In an image that is simultaneously biological and profoundly abiological, Tzara then portrays the poet as someone who knows the tree that he planted, but not what fruits it will bear.[5] Poetry becomes a preeminent "medium of knowledge"[6] precisely at the point when it transcends its function as a carrier of communication and meaning and begins instead to take itself at its own word.

In a letter to the young publisher Roger Caillois, Bachelard explained what it was about surrealism that fascinated him as a philosopher of science and guided his thinking: "The surrealists liberate us from dogmatic rhythms. When reading them, I often rediscover a temporal agility that has been crushed by the ponderous weight of philosophy. One must make use of this liberation in order to achieve an aesthetics of the abstract."[7] Reviewing the *Inquisitions* issue for *La nouvelle revue française*, Jean Wahl located Bachelard in the tradition of Léon Brunschvicg, particularly pointing out the close proximity between Bachelard's surrationalism and the most recent developments in science.[8]

At the very least, Tzara's description of experimental poetry's self-transcendence as an interaction between "directed thinking" and "undirected thinking"[9] must have forcefully reminded Bachelard of his own attempts to describe the experimental excess arising from work on scientific objects—a transcendence in which one took oneself, not at one's words, but at one's things, as it were. In the closing chapter of *The New Scientific Spirit*, proposing a "non-Cartesian epistemology," Bachelard articulated this as follows: "When, moreover, we realize what an incomplete state modern science is in, we begin to gain some intimate idea of the meaning of 'open-minded rationalism' (*le rationalisme ouvert*)."[10] He then asks, fully in line with the figure of excess: "If we really want to understand our intellectual evolution, wouldn't we do better . . . to pay heed to the anxiety of thought, to its quest for an object, to its search for dialectical opportunities to escape from itself, for opportunities to burst free of its own limits? In a word, wouldn't we do better to focus on thought in the process of objectification?"[11] It is surely significant that Bachelard, looking at its trace-generating instruments, characterizes contemporary physics as an epigraphy of matter—a "new way of 'reading' matter"[12]—and thus deliberately places it in the purview of written script.

In Bachelard's encounter with Albert Flocon's techniques of perspective, this motif of a non-Cartesian epistemology would acquire a fresh dimension and a new urgency. In fact, Bachelard himself returned to the theme in a literary context, as part of his interest in Paul Éluard's poetry. It was Éluard who, in 1948, had paved the way for Flocon's first book of prints, accompanying it with some of his poems. And it was lines by Éluard that opened and closed Bachelard's *The Psychoanalysis of Fire* in 1938. His 1953 homage to Éluard, entitled "Germe et raison," began with the following words: "Germ and Reason—the two poles of the poet's immortality. Reborn through the germ, he lives on through reason."[13] For Bachelard, Éluard's poetry embodied the relationship between novelty and tradition in all their reciprocal provocation.

One might be tempted to interpret Bachelard's collaboration with an engraver as a philosopher entering into the oeuvre of a visual artist whose labors epitomize the chiasmus of reason and dream quoted earlier: a surrational organization of the real and a surreal organization of the rational, in one and the same movement. In this work—in a graphical world shimmering between spatialization and despatialization, figuration and defiguration, concretion and abstraction—Bachelard both recognized the tenor of his surrationalist manifesto and was able to retrace and reexperience it as an act of audacity: "There should be no hesitation: one should choose the side where one thinks the most, where one experiments the most artificially, where ideas are the least viscous, where reason loves to be in danger. *If, in any experiment, one does not risk one's reason, that experiment is not worthwhile attempting.*"[14]

ALBERT FLOCON, GASTON BACHELARD

Albert Flocon was born Albert Mentzel in 1909, into an engineer's family in Köpenick, near Berlin. He grew up in the Saxon town of Döbeln and later attended a progressive boarding school in Haubinda, Thuringia. In 1927, Mentzel began to study at the Bauhaus in Dessau, hoping to become an architect. As an eighteen-year-old student, he was introduced to the world of Bauhaus in the famous "Vorkurs" offered by painter and art theorist Josef Albers. Describing this experience in his memoirs, Flocon wrote: "The aim of the course was not to produce objets d'art, but to carry out research into the formal logic of materials, to test out their techno-aesthetic possibilities."[1] The students sought to unmask the traditional image of the artist and "enter into direct contact with the materials."[2] These were not precious substances—they might just as well be a sack of empty matchboxes, or tin cans, or iron filings. The introduction to theories of point, line, and plane was given by Wassily Kandinsky.[3] After meeting the painter and stage designer Oskar Schlemmer, Flocon abandoned his architectural ambitions. He turned to theater and was soon participating in guest performances at the Bauhaus theater.[4] Schlemmer's analytical view of the stage had a lasting influence on Flocon's artistic and scientific development,[5] as is indicated by his late tribute to Schlemmer in *Scénographies au Bauhaus*.[6] There was one more ingredient of the Dessau Bauhaus that stayed with Flocon: "The Dadaist elements of surprise and disconcertment, also to be found in Schlemmer's work, were the foundation of everything that was said, written, and created at the Bauhaus."[7]

In 1930, Albert Mentzel left Dessau and moved to Berlin. In view of his markedly left-wing political views and his marriage in the early 1930s to a woman of Jewish descent, Lotte Rothschild of the Frankfurt manufacturing family Rothschild, he decided to seek sanctuary in France with his wife and young daughter Ruth in 1933. In Paris, Mentzel initially worked in advertising, with mixed success, but he also tried his hand at various book projects. These ranged from a volume of nude photographs, including pictures by Man Ray, to children's books for the popular "Père Castor" series, published by Flammarion under the aegis of Paul Faucher.[8] From 1936, Mentzel found work with Victor Vasarely, who had moved to France from Hungary in 1930 and at this time ran a studio in Montrouge, near Paris, that specialized in pharmaceuticals advertising. When war broke out in 1939, Mentzel was interned as a German citizen. To avoid being sent back to Germany, he joined the Foreign Legion and served in Algeria until 1941. After demobilization, he first stayed in Pibrac, near Toulouse, with the *Annales* writer Lucie Varga, whom he had met in the late 1930s in the Paris suburb of Viroflay. His family joined him in Pibrac, and he managed to make a living from occasional painting commissions. After the occupation of the south of France and shortly before liberation, he was arrested by agents of the *police de l'air*—his wife was working at Sud-Aviation.[9] She and their eldest daughter were deported on one of the last trains from Drancy and were killed in Auschwitz. Mentzel survived the war in the Toulouse panopticon of Saint-Michel, his two younger children Catherine and Henri with acquaintances in the countryside. Albert Mentzel returned to Paris in late 1945. He adopted the name of a revolutionary French ancestor, Ferdinand Flocon,[10] and became a French citizen.

Gaston Louis Pierre Bachelard was born in 1884 in Bar-sur-Aube, Champagne.[11] His father ran a tobacco store; his grandfather had been a shoemaker. Bachelard would later dream of his province thus: "I was born in a section of Champagne noted for its streams, its rivers, and its valleys—in Vallage.... The most beautiful of retreats for me would be down in a valley, beside running water, in the scanty shade of the willows." He owed his "fundamental color," he wrote, to the landscape of his birth.[12] After his school education and a degree in mathematics and physics at the Sorbonne, Bachelard worked for the postal service in Paris until the outbreak of World War I. From 1913, he prepared for training as a telegraph engineer, but he just missed selection for the course, and wartime mobilization prevented him from trying again. Bachelard spent the whole of the war, from August 1914 to March 1919, as a soldier in action. His wife, Jeanne Rossi, died in 1920, when their daughter Suzanne was not yet a year old; he went on to bring her up alone. After the war, he taught physics and chemistry at the high school in his home village of Bar-sur-Aube. In parallel, he also studied and taught philosophy—"without having attended a single lecture," as he remarked in a letter to Gustave Le Bon on his "rather irregular intellectual life."[13] Bachelard took his doctorate in philosophy of science and history of science at the Sorbonne in 1927 under the guidance of Abel Rey and Léon Brunschvicg, writing on approximate knowledge and the history of thermal propagation in solids.[14] He taught from 1927 to 1930 as a lecturer and then, from 1930 to 1940, as full professor at the University of Dijon, succeeding Georges Davy. In Dijon he wrote, in quick succession, the epistemological books that earned him his reputation not only as a historical epistemologist, but as a thinker interested in the constant forward movement of the sciences. These works included *The New Scientific Spirit* and *The Formation of the Scientific Mind*,[15] along with the first part of his tetralogy on the imagination of the elements, *The Psychoanalysis of Fire*.[16]

In 1940, after some hesitation, Bachelard agreed to take the place of his teacher, Abel Rey, as professor of history and philosophy of the sciences at the Sorbonne and director of the Institute of Philosophy and History of Science and Technology. As he admitted to his wartime friend Marius Filloux: "This is the drama that has been agitating me for ten days. Here [in Dijon] I have a dream life. We love the house, where we have lived for a year. At the faculty, I have peace and quiet, and very pleasant working conditions. . . . [But] I have been given to understand that this will be my last chance to go to the Sorbonne."[17] After publishing a further epistemological work the same year, *The Philosophy of No*,[18] alongside his teaching in the philosophy and history of science in Paris he subsequently concentrated on completing his essays on the four elements of fire, water, earth, and air.[19] They grew from his observations on how these elemental figures were realized in the literary imagination as images of motion and rest, of forces and resistances. During his decades at the Faculté des Lettres in Dijon and in the Parisian Latin Quarter, Bachelard also maintained a wide spectrum of contacts with the French literary and artistic avant-garde, especially the surrealists and their affiliates. Even after he moved to Paris, he continued to spend his summers in Dijon, in Burgundy near his home region of Champagne, where he worked on his books in seclusion.

PERSPECTIVES

Flocon began his career as a copperplate engraver after returning to Paris in late 1945. During his imprisonment in Toulouse, he had become interested in perspective drawings—"something like an exploration of space" under claustrophobic conditions.[1] He continued these studies in Paris. Flocon's American friend Henri Goetz, a graphic artist who had also been living in Paris since the 1930s and participated in the wartime Resistance, put him in touch with the painter and engraver Georges Visat. It was Visat who convinced Flocon that engraving on copper would be an ideal medium for realizing his ideas. He also accompanied Flocon's early experiments in copper engraving. The first of these showed "an open hand pointing in one direction, in the palm of which an eye opens, in other words a hand that sees, the hand of the engraver"[2]—and, one might add, a hand that is still a work in progress, as yet missing three fingers (fig. 1a). A surviving proof (fig. 1b) reveals how the geometrical frame, the sheer austerity of which gives the hand its spatiality, altered during its graphic implementation. Juxtaposing the two images also highlights the constraints upon all additional interventions into an engraving once it has been started. A stroke that has been drawn cannot be undrawn.

Flocon placed this image at the beginning of his first album, *Perspectives*, which was printed at the Imprimerie Union, engraved at the Atelier Georges Visat, rue de Bourbon-le-Château, and published by the art publisher Aimé Maeght, who had opened a Paris gallery in 1946 and founded the art periodical *Derrière le miroir*.

Accompanying the ten *Perspectives* engravings were ten short poems by Paul Éluard.[3] Flocon had met Éluard through the bookseller Lucien Scheler, who, himself an author and active in the Resistance, knew the poet very well—he had sheltered him from 1942 to 1944. The contact originally derived from the publisher Georges Blaizot, who had advised Flocon to bring in a famous name for his debut publication rather than writing the accompanying texts himself.[4] Éluard wrote the following four lines for the hand image:

Je noue et je délie je donne et je refuse
Je crée et je détruis j'adore et je punis
Ma fleur est la pensée je caresse et je sème
Je vois avec mes doigts je touche et je comprends.[5]

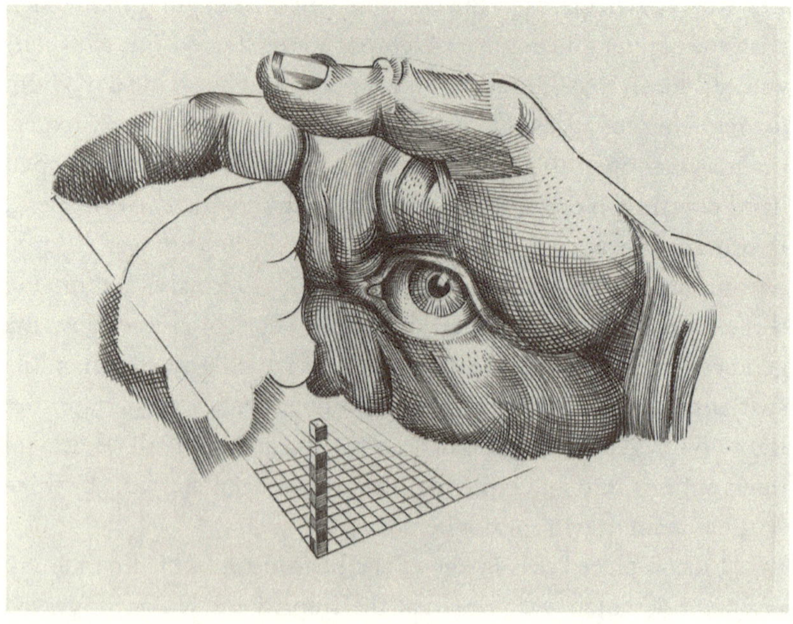

Figure 1a. Plate 1 (15 x 20 cm) in Paul Éluard and Albert Flocon, *Perspectives: Poèmes sur des gravures de Albert Flocon* (Paris: Maeght, 1948), copy no. 81 (of 150 on vélin de Lana; total edition 200).

I knot and I untie I give and I deny
I create and I destroy I worship and I punish
Thought is my flower I sow and I caress
I see with my fingers I touch and I grasp.

Éluard's lines distill the ambivalence or even antinomy of the hand's possible gestures, and simultaneously the generative and cognitive force of its tactile operations.

In their poetic potency, these four lines are in no way inferior to the prosaic concision of the panegyric "In Praise of Hands" that Henri Focillon added to his major art historical study *La vie des formes* when its new edition came out in 1939.[6] This was the same year that he met Bachelard at the last of the annual Décades de Pontigny symposia.[7] "The hand," runs one of Focillon's rhapsodic sentences there, "means action: it grasps, it creates, at times it would even seem to

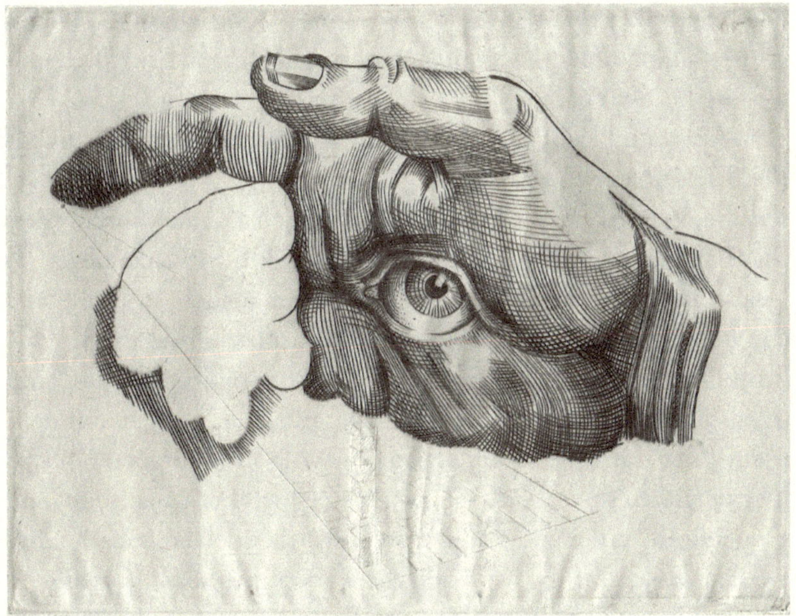

Figure 1b. Proof of 1a, single page.

think."[8] Then the last note of conjecture disappears: "Art is made by the hands. They are the instrument of creation, but even before that they are an organ of knowledge."[9] The art historian Focillon, who was also a poet and engraver, died in exile in the United States in 1943. He and Éluard belonged to two different generations, but their contemplations of the hand are united by a shared proximity to the visual arts. It is no coincidence that André Breton described Éluard, an ally from his early surrealist days, as "the painters' friend." Perhaps more of a coincidence is the fact that texts by Éluard and Focillon can be found in print together after Focillon's death, for example, in the new journal *La table ronde*, established in Paris in 1944.[10]

The theme of the seeing and thinking hand was nothing new; the "intellectual supercharging of the hand" pervaded surrealist art from the outset.[11] An example is *Les mains libres*, the joint work of Man Ray and Paul Éluard created in 1936 and 1937, with its title, its cover image of hands designed by Jeanne Bucher, and the thematic centrality that it gives to the gestures of the hand.[12] Neither is it surprising that Dorka Raynor chose to photograph Éluard's hands writing his own name.[13]

Any number of further examples could be cited. Just one is the Swiss writer Denis de Rougemont's 1936 diatribe against intellectuals, *Penser avec les mains*: "It is said that some think, others act! But the true condition of man is to think with his hands. . . . I call 'hand' that which *manifests* thought, renders it visible and corporeal; which renders it, in the double sense of the word, *grave* [serious; pregnant]."[14] In his expressionist variation on the panegyric to the hand, however, the disputatious de Rougemont does not refer to the same hand that was enjoying such burgeoning popularity within surrealism, the milieu in which both Bachelard and Flocon were at home. For the surrealists, the social distinction between the work of the hand and the work of the head, so zealously attacked by de Rougemont, seemed to have been largely overcome in the experimental sciences and the arts. In both universes, the experimenter was unthinkable without his hands, feeling their way and making their interventions. In Focillon's work, too, the parallel between artistic craft and scientific experimentation is impossible to miss. On tools, he writes: "Anyone who has not known men who live by their hands cannot understand the strength of these hidden relationships" between hand and implement, in which is found, "on the highest plane, the concern for experimentation."[15]

A surrealist periodical published clandestinely between 1941 and 1944 took the name *La main à plume*, alluding to the famous line by Arthur Rimbaud.[16] The group's members regarded their writing as political action, and many—including Goetz and Éluard—played an active part in the Resistance. Not least, the hand was a leitmotif in the Bauhaus school. "Albers told us," recalls Flocon, "that one must see with one's hands. To be able to do something with the materials, one must touch them, feel them, test them."[17] I will stop there. I am not concerned here to track the historical florescence of the hand motif in all its ramifications and its rich diversity of contexts. My subject, rather, is the specific form that it assumed in the encounter between the writer Bachelard and the engraver Flocon.

If one hand opens Flocon and Éluard's *Perspectives* collection, another hand closes it (fig. 2).[18] In this case, it is the hand of the picture framer, the profession that Flocon followed for a while to support his family during the war. Almost life-size, the hand takes on a life of its own, while the comparatively small body is squeezed into a cramped rectangle that it tries to split open and must nevertheless carry. The prominent eye of the oversized face in the background, detached from the frame, gazes suspiciously at the liberated hand stretching out before it. In the accompanying poem, Éluard writes of a "glittering creatrix," leaving it open whether this refers to the hand, the eye, or the Promethean figure braced within the quadrangle. All spatial confinement here seems disrupted, burst apart.

Flocon's memoirs make written play with the hand motif on several occasions, too. "To grasp intellectually is also to take, it is the hand that takes, the hand, universal pincer, hammer descending, lever lifting the world. For me, every apprenticeship has been linked with the hand that must be implicated and the eye that verifies, watches, commands, judges, weighs, and weighs up."[19] The early hands in *Perspectives* contain a germ of reflexivity in the form of a self-referentiality that is inscribed in the process of fabrication. This was to become a hallmark of Flocon's oeuvre and would soon attract the interest of Gaston Bachelard.

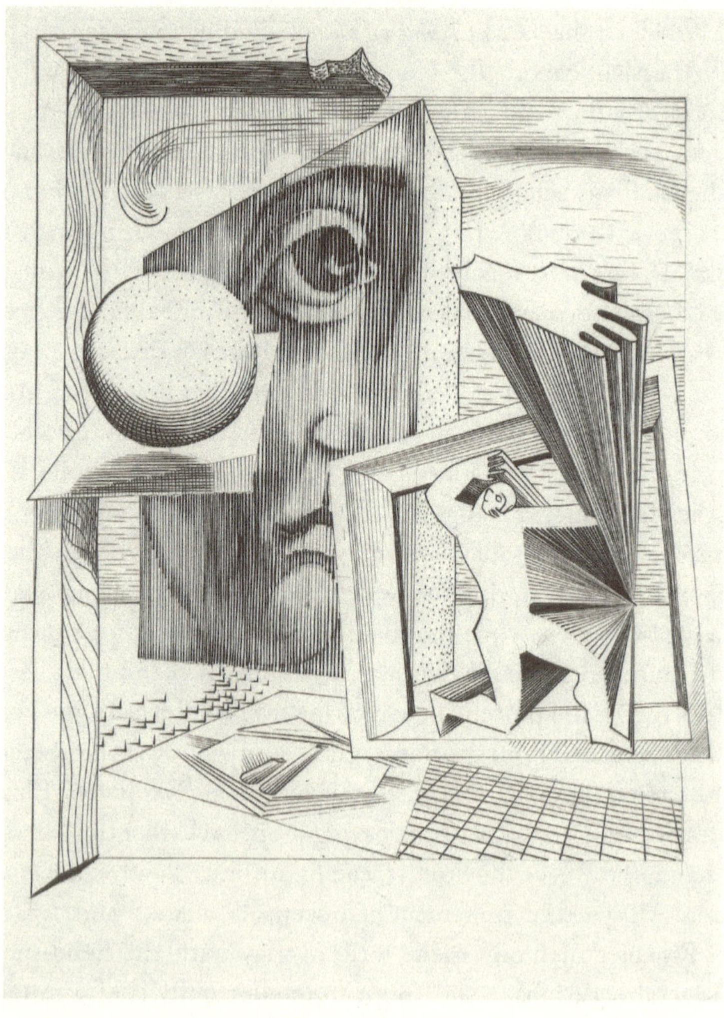

Figure 2. Plate 10 (24 x 18 cm) in Éluard and Flocon, *Perspectives.*

REVERIES OF EARTH

In 1948, Gaston Bachelard published the final volume of his tetralogy around the four elements: the volume on earth, which appeared in two parts with José Corti.[1] *The Psychoanalysis of Fire* (1938) had been followed in 1942 and 1943 by the volumes on water and air.[2] The last book that Bachelard published in his lifetime returned to a particular form of fire, *The Flame of a Candle*,[3] and he was working on a poetics of fire when he died in 1962.[4] Bachelard's concern with the imagination of the "fundamental material elements," as he once called them,[5] thus stretched over a quarter of a century.

In the first part of the earth study, *Earth and Reveries of Will*, Bachelard discusses images of earthen matter and the ways he has seen them handled in his extensive reading of poetry and fiction, especially that of the nineteenth and twentieth century. His point of departure is the central claim that "Earth, unlike the other three elements, is first and foremost characterized by *resistance*."[6] In the very first chapter, Bachelard addresses the gesture or "act" of labor, which "brings integration not only with the object that resists, but also with the essential resistance of matter itself."[7] At the two poles of this gesture are matter and the hand: "Hand and Matter must become one in order to form the point of intersection for this *energetic dualism*, an active dualism quite different from the classic dualism of object and subject."[8] The hand forms part of the subject and the object at once: "By means of such images of work upon matter, workers acquire such a fine appreciation of the qualities of different materials and develop such an intimate relationship with their material values that one could say they come to know these substances genetically, as if they might bear witness to their fidelity to the basic structures of matter."[9]

At the heart of the chapter on hard materials and the tools for processing them is what Bachelard calls a "lovely metaphysical couplet" by Paul Éluard on polishing, that "strange transaction between subject and object."[10] The couplet is difficult to translate, flashing iridescently between grammatical subject and object but nevertheless as monumental as the stone that is its theme:

Cendres, polissez la pierre
Qui polit le doigt studieux.

Ash, polish the stone
That polishes the studious touch.[11]

Ash, the polishing agent, is the subject of the first line. Its object, the stone, becomes the subject of the second line, the object of which—the finger—works the whole.

This image conveys another message as well. The tool itself becomes ambivalent, both chafing the object and clinging to it. This dual character of the tool, assailing and apprehending, contains in condensed form the two sides of the gesture that epitomizes work. In the foreword to the second *Earth* volume, Bachelard notes that whereas the first, subtitled *Essai sur l'imagination des forces*, was written from the perspective of the preposition *against*, this complementary *Essai sur l'imagination de l'intimité* pursues "the preposition *in*."[12] As Éluard's couplet illustrates so well, the two perspectives are ultimately inseparable. The most powerful "material images of earth," Bachelard observes, realize "an *ambivalent* synthesis in action, uniting *against* and *in* dialectically."[13]

IN PRAISE OF HANDS

Around the time when Bachelard's *Earth and Reveries of Will* and *Earth and Reveries of Repose* were published, Flocon and some like-minded Parisian colleagues founded "Graphies"—a group of "true engravers" who printed their own designs rather than entrusting them to external printers. Flocon became the secretary of this otherwise very loosely organized group. With a Swiss colleague, Albert-Edgar Yersin, whose acquaintance he had made in the early days of his new profession in Paris and whom he called "the best copper engraver in the world,"[1] Flocon began to organize an exhibition for the group in the Galerie des Deux Îles, a small establishment on the Île Saint-Louis that no longer exists today. For the presentation of the exhibition, he considered a "montage of quotations" from *Earth and Reveries of Will.* Flocon had only recently become aware of Bachelard's books on the four elements and their worlds of language.[2] He was particularly fascinated by the reveries on earth, which resonated profoundly with the engraver's work. According to his memoirs, when he asked Bachelard's publisher José Corti for permission to cite the works, Corti suggested that he approach Bachelard directly instead. Flocon wrote to the philosopher, and the two met in Bachelard's apartment on rue de la Montagne Sainte-Geneviève. Bachelard had Flocon explain the graphic techniques used in the individual plates and was taken with Flocon's plan, but he objected that one should "never print the same thing twice."[3] He promised instead to write something new for Flocon's project. Bachelard's "Matière et main," excerpts of which were enclosed in the exhibition invitation, appeared as the introduction to the album *À la gloire de la main*, which brought together the prints shown in the exhibition.[4] The phrase that Bachelard originally chose for his short essay had now become the title of the work as a whole.

Flocon, who had made contact with the group around André Breton soon after the war,[5] regarded the book as a kind of manifesto for an art that "considers itself more of a practice than an aesthetics," a practice that moves "indefinably between surrealism and geometrical or lyrical abstraction."[6] At its core was "man grappling with matter," wrote Flocon in the foreword as the spokesman of the Graphies group, and the album was intended to "signal some of the grips that are possible for the hand as it feels its way forward." The determination to pursue a particular path and the need first to open that path up, to sense its course through experience, are inextricably united in this work. It is telling that Flocon here addresses engravers as "artisans," not as artists. He reiterated the point in a later work: "I felt and still feel a profound discordance with the 'spiritual in art,' the infinite variations of which serve to camouflage platitudes. Art, if it exists, seemed and still seems to me to be a material activity. The spirit comes upon us as an extra."[7]

À la gloire de la main contains sixteen prints by the group's artists on the theme of hands, preceded by a number of texts, of which Bachelard's is the first. The book was printed in 1949 in the famous studio of Georges Leblanc at 147, rue Saint-Jacques, "at the expense of an art lover." Flocon's contribution (fig. 3) shows his own hand, entitled simply "My Hand." It is holding a burin and is itself in the process of engraving a hand—a hand, then, "at work" on a hand.[8] The hand being drawn continues the chain of defiguration and transfiguration: from what might be called the physical hand, which is incising the outline of another hand into the copper, in the palm of which yet another hand can be seen, now in the abstracted form of a geometrical sketch. The picture shows something else as well: the gesture of the engraver simultaneously incises and

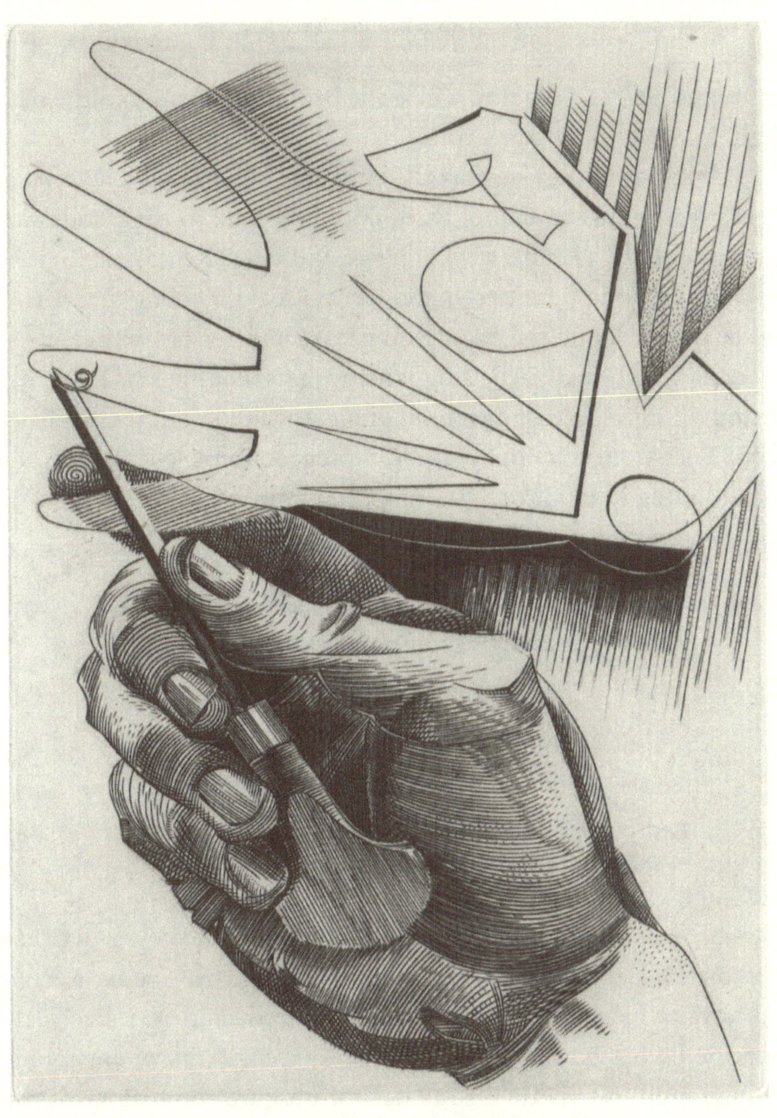

Figure 3. Plate 7 (17.5 x 12.5 cm) in Gaston Bachelard, Paul Éluard, Jean Lescure, Henri Mondor, Francis Ponge, René de Solier, Tristan Tzara, and Paul Valéry; engravings by Christine Boumeester, Roger Chastel, Pierre Courtin, Sylvain Durand, Jean Fautrier, Marcel Fiorini, Albert Flocon, Henri Goetz, Léon Prébandier, Germaine Richier, Jean Signovert, Raoul Ubac, Roger Vieillard, Jacques Villon, Gérard Vulliamy, and Albert-Edgar Yersin, *À la gloire de la main* (Paris: Aux dépens d'un amateur, 1949), copy no. 67 (on vergé d'Arches) of 164.

excavates. It proceeds down into the bedrock and up into the daylight at one and the same time.

Looking at the list of artists represented in *À la gloire de la main*, it is striking that several of them once belonged to the clandestine group around *La main à plume*: Gérard Vulliamy, Raul Ubac, and the couple Christine Boumeester and Henri Goetz. Among the writers, we again encounter the poets Paul Éluard and Tristan Tzara, whose work had interested Bachelard for many years. For Flocon, Tzara may have awakened dadaist memories of his student days at the Bauhaus. Four lines from a poem by Tzara, arranged around the picture of a hand setting in motion a "frail language," suggest as much:

tu chantes des berceuses dans la langue de ta lumière
aux fraîcheurs des nuits presbytes enveloppées dans les langes de vent
mère des chansons égorgées dans la vague noyées
ta main sait agiter tant de langage frêle[9]

you sing lullabies in the tongue of your light
in the cool of wind-swaddled farsighted nights
mother of throat-slit songs drowned in the wave
what a frail language your hand can stir up

This publication marks the first personal encounter between Flocon and Bachelard—an encounter that would become an "elective affinity," as Flocon later called it.[10] In his introductory text, Bachelard specified what he saw as the emblem or peculiarity of engraving: "The happily aesthetic results do not obscure the history of toil that lies behind them, the history of the battle with matter." He was drawn to the continued presence of the productive process in the product itself, where "awareness of the hand at work comes alive within us as a sharing in the engraver's craft."[11] Two things about the task of engraving particularly fascinated Bachelard and prompted him to comment again and again during his engagement with Flocon's oeuvre: the forms of resistance in the confrontation with matter; and the hand as the material and knowing agent of a becoming, of an emergence. These two strands are inseparable. Resistance is what elicits becoming, what challenges and imprints itself on becoming—"becoming" not as a passive event but as a garnering, a provocative examination. Paul Valéry concluded the text he wrote for *À la gloire*, and with it the textual section as a whole, with the following complex question:

> How shall we find a formula for this apparatus that by turns strikes and blesses, receives and gives, nourishes, takes its oath, sets its tempo, reads for the blind and speaks for the mute, stretches out to the friend, rises up against the enemy, that becomes a hammer, becomes pincers, becomes alphabet? . . . Who knows? This almost lyrical disorder suffices. Successively instrumental, symbolic, speaking, and calculating—a universal agent, might one not call it an *organ of the possible*, just as it is also *the organ of positive certainty*?[12]

Stefan Bollmann notes that Bachelard glimpsed "the gesture of engraving itself" in the plates—"this feat, in the multiplicity of its possible expressions, recapitulates the feat that for human beings signifies the appropriation of the world."[13] "Every hand is an awareness of action," as Bachelard remarks.[14] But this is no mere feat of strength; it is with "utter delicacy" that the hand evokes "the prodigious forces of matter! All dynamic dreams, from the most violent to the most insidious, from the metallic furrow to the finest of lines, are there in the human hand, synthesis of strength and skill. Hence the variety and at the same time the unity of a volume in which sixteen great workmen each lay before us the life of a hand."[15]

As Flocon recalls in his memoirs, Bachelard was "always curious about the tools, the manipulations, the material labor."[16] Not only did he peruse the engravings themselves with care, he also asked to visit a workshop, and at Leblanc's studio, in the second courtyard of 147, rue Saint-Jacques, had the whole array of tools explained to him in great detail.[17] More generally, Flocon reports, he "always asked people about their craft"[18] and wanted to hear everyone talk about "what he knew about best."[19] It was in Leblanc's spacious studio that Flocon and the graphic artist and etcher Johnny Friedländer, whom he had also met at Leblanc's, set up the "Atelier de l'Ermitage" in 1950. This free school of graphics taught etching and engraving at the same location until 1954, enabling Flocon to devote himself continuously to copper engraving and wrest new dimensions from the technique, which at the time seemed in danger of falling into oblivion. These innovations came to fruition most obviously in Flocon's much later "experimental series," in which he subjected his prints to further treatment, experimenting especially with set pieces and alternative printing materials. This gave the engravings a second life.[20]

Looking back, Flocon described the experimental prints as follows: "Starting from the very first plates for *Perspectives*, I performed experiments . . . by carrying out unorthodox prints to see what would happen. The lovely austerity of the burin's finished line did not seem to me to be its final state. In an intaglio studio, there are too many materials, too many instruments lying around, quite apart from the presses; there are too many manipulations and dexterous tricks of the hand for a craftsman to be able to resist appropriating this rich knowledge and immediately channeling the skills he has gained into new directions."[21] Flocon noted elsewhere: "A simple technique such as intaglio is, by its very essence, rich in possibilities: just devoting the necessary time to experimentation is enough for it to become pliable, give way, and produce unexpected results."[22] In his book on prints, Georges Didi-Huberman likewise stresses that the "gesture" of the print "does not intrinsically possess the targeted and utilitarian value of the production of an object: it is above all the *experience* and *experiment of a relationship*." From this, Didi-Huberman derives a "principle" of the print that in fact amounts to "the following non-principle: *one never knows exactly what will ensue from it.* In the process of the print, form is never completely 'fore-seeable': it is always problematic, unexpected, unstable, *open*."[23]

LANDSCAPES

March 1950 saw the publication of *Paysages*, a substantial volume with sixteen of Flocon's engravings, for which Bachelard wrote the introduction and extensive texts to accompany the plates.[1] In April, *Cobra*, the journal of a group of experimental artists around Constant Nieuwenhuys and Asger Jorn, called attention to the work by publishing some extracts, originally intended as a preprint.[2] Flocon structured the sequence of images as a play on the theme of metamorphosis: "I had begun a series of engravings on the theme of metamorphoses, correspondences; there were human bodies that are landscapes one traverses, anthropomorphic landscapes, an exploration, a geography from the field to the desert or the plain to the mountain, from the head to the foot, between the two poles both of a human body and of the Earth."[3] At the same time, it engaged with the four elements, "inspired by reading Bachelard's books on the imagination of the elements: fire, air, earth, and water."[4] Flocon was attracted by the challenge of complementing Bachelard's poetics of the elements with a metamorphic graphics of the elements, in which "fire is represented at once by flames and by a couple taking flight, water by an eye that is also a lake, earth by a female body that is worked like a field."[5]

In summer 1949, a minor but telling dispute arose between artist and writer regarding the title of the book. Flocon originally suggested "Du paysage."[6] From his summer residence in Dijon, Bachelard informed Flocon that he would prefer the book title to be the previous provisional title of his own introduction, "Le paysage gravé"; Flocon could then choose a different subtitle himself.[7] Bachelard thought Flocon's title would be detrimental to the book's sales.[8] This worry may seem surprising, but it soon becomes clear that Bachelard thought of his publications on a very different scale, and he wanted at least to achieve the maximum reach possible for a book that would appear in such small numbers. When he heard that the book, and thus his texts, would have a print run of merely two hundred, Flocon recalls, he was "dismayed."[9] In the end, the title *Paysages* was agreed upon,[10] and Bachelard called his prefatory text "Introduction à la dynamique du paysage."[11]

Bachelard read the "landscapes" of *Paysages* above all as manifestations of movement. He introduced the book thus: "In losing color—the most potent of all sensual attractions—the engraver retains one great opportunity: he can and must discover *movement*. Form is not enough. A passive copying of form alone would make the engraver nothing more than a lesser painter. But in the forceful world of engraving, line is never mere profile, sluggish outline, arrested form. The least line in an engraving is a trajectory, a movement."[12] At the end of his introduction, Bachelard comments that an engraving is accompanied by "a kind of immediate gladness without conscious cause," which is what defines the engraver's "taking possession of the world." He concludes: "The engraver's landscape is an act"[13]—it is a transgression.

Pascal Fulacher has described Bachelard's texts for this volume as "prose poems";[14] Bachelard himself more soberly entitled them "notes by a philosopher for an engraver."[15] In his memoirs, Flocon remembers their endeavor: "I brought Bachelard one or two plates a fortnight, so I saw him often and our friendship unfolded out of this shared work, in which the images are anterior to the text that illustrates them."[16] Even if the term "illustrate" does not seem quite the right one for Bachelard's texts, this reversal of the conventional relationship between text and image—the one typical of illustrated books—highlights the special nature of the correlation here. What occurred in this interplay was the creation of two parallel universes, one consisting of words and the other of incised lines. Flocon believed that the texts "can very well stand alone, without the support of the image of which they form part. In any case, there would, I think, be a very fine dissertation to be written on the relationship and the irreducibility of a verbal image and an image-object, an engraved image."[17] As Bachelard put it: "These notes on Albert Flocon's engravings are the reactions of a solitary philosopher. If they have no other virtue, they have at least the spontaneity of the reclusive dreamer. In the case of the majority of the plates, Flocon and I did not look for any 'common ground.' "[18] No communication, then, in the usual sense of the word—*pas de discours*. Yet Flocon is right to say that "the images are anterior to the text"; after all, Bachelard "reacted" to them. For Bachelard, "if the poet's landscape is a state of mind, the engraver's landscape is a disposition or outburst of will, an activity that is impatient to come to grips with the world. The engraver sets a world in motion, . . . provoking the forces that lie dormant in a flat universe. Provocation is his way of creating."[19] The engraver's labor is an object lesson in "work with hard materials," as Jean Starobinski put it in his review of the volume for *Journal de Genève*.[20] This was one of the few written responses elicited by the remarkable book, which was printed by Fequet and Baudier in Paris, engraved at Georges Leblanc's studio, and published by Paul Eynard in Rolle, a small Swiss town in the Canton of Vaud. Encouraged by Yersin, Eynard had also advanced the funding for the whole enterprise.[21]

Let us venture a few glimpses into *Paysages*, beginning with what I will call Flocon's self-portrait (fig. 4a). Preceding the plates as a kind of epigraph, this unnumbered and untitled image takes up the thread of the fractured self-portrait that closed the *Perspectives* collection. Flocon elsewhere referred to it as the "head of a Minerva under construction."[22] Minerva is the protectress of craftsmen, poets, and the wise—and in this image, the emblem of work itself becomes

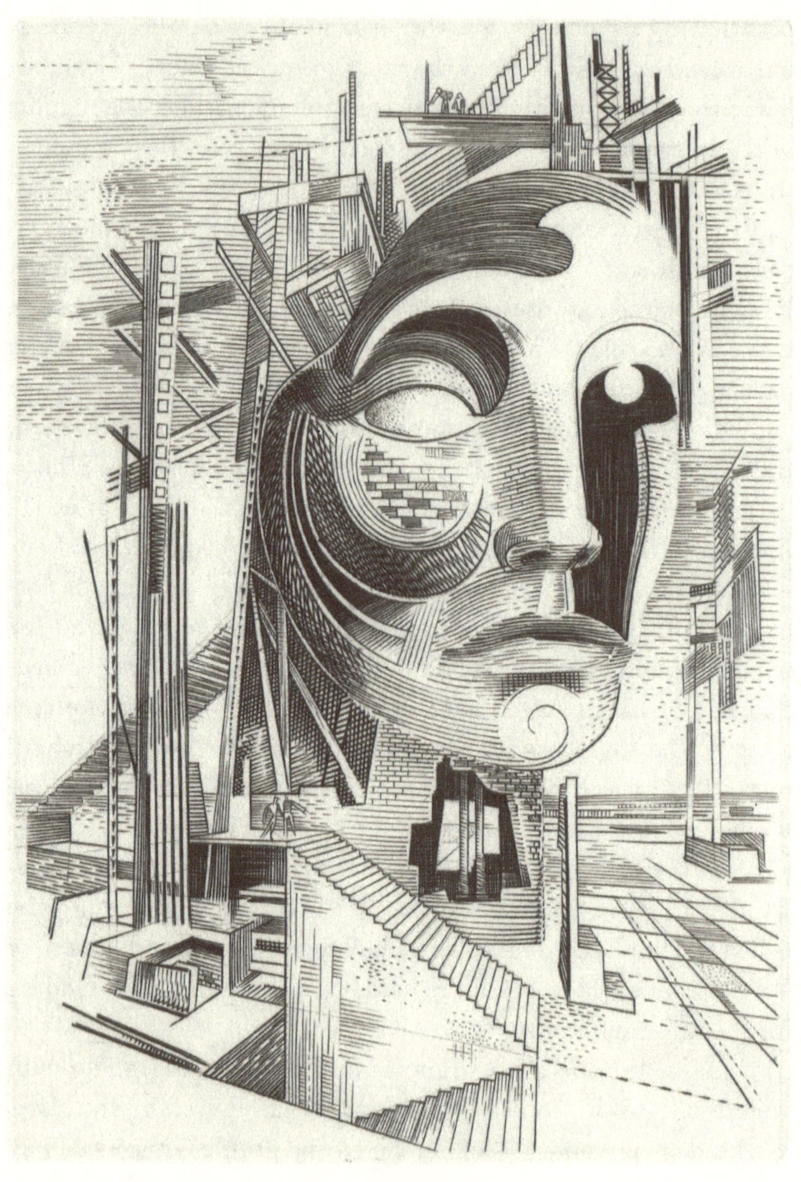

Figure 4a. Frontispiece (18 x 12.5 cm), in Gaston Bachelard and Albert Flocon, *Paysages* (Rolle: Paul Eynard, 1950), 4, copy no. 7 (on vergé d'Arches) of 220.

a construction site. A preliminary pencil sketch of the plate survives, in reverse compared to the print (fig. 4b), and there is also a later drawing in ink (fig. 4c). The ink drawing is dated, proving that it was made later than the print.[23] Thanks to these drawings, it is possible at least partially to reconstruct the engraver's game. That game has its own momentum, as a comparison of the different images reveals. The way the engraver guides the burin, the way he is obliged to follow it, differs from the way the draftsman guides the pencil or moves the pen.

The image shows a head in the process of being built, set in a flat and barely delineated landscape. It is impossible to tell whether the scaffolding we see is supporting the head's construction or growing rampantly beyond the head, appropriating it as a component of a larger architecture whose dominating vertical struts contrast conspicuously with the rounded lines of the face. In a later interview, Flocon compared this vista with the bleak summit of Mont Chauve, north of Nice, which he remembered from his time in the *zone libre*.[24] At first sight, the drawings and the print are very similar, although the print creates a far heavier, more powerful impression. All three images make it immediately obvious what Bachelard meant when he said that the engraver's landscape was an "act," a deed. Yet once we begin to attend to details, it soon becomes clear that very few correspond exactly between the drawings and the print. In the print, the right cheekbone is filled with bricks; in the ink drawing, the bricks mark the temple and the bridge of the nose; and in the sketch, they are completely absent. Where the print offers a view into the construction of the building's cavity, the drawing shifts the focus to its outer framework; in the sketch, neither impression is present. Walkers populate the path around the site in the drawing, whereas in the print the path is replaced by a checkered forecourt, which in the sketch is blank. Perhaps the most striking difference is the staircase. In the drawing, it responds with a countercurve to the rounding of the face; in the engraving, it has hardened into a rigorously linear flight of diagonal steps down which two figures seem about to descend. The detail emphasizes the sense of the engraver's provocation that Bachelard names in his introduction: "Provocation is his way of creating."[25] Finally, in the engraving another stairway, above

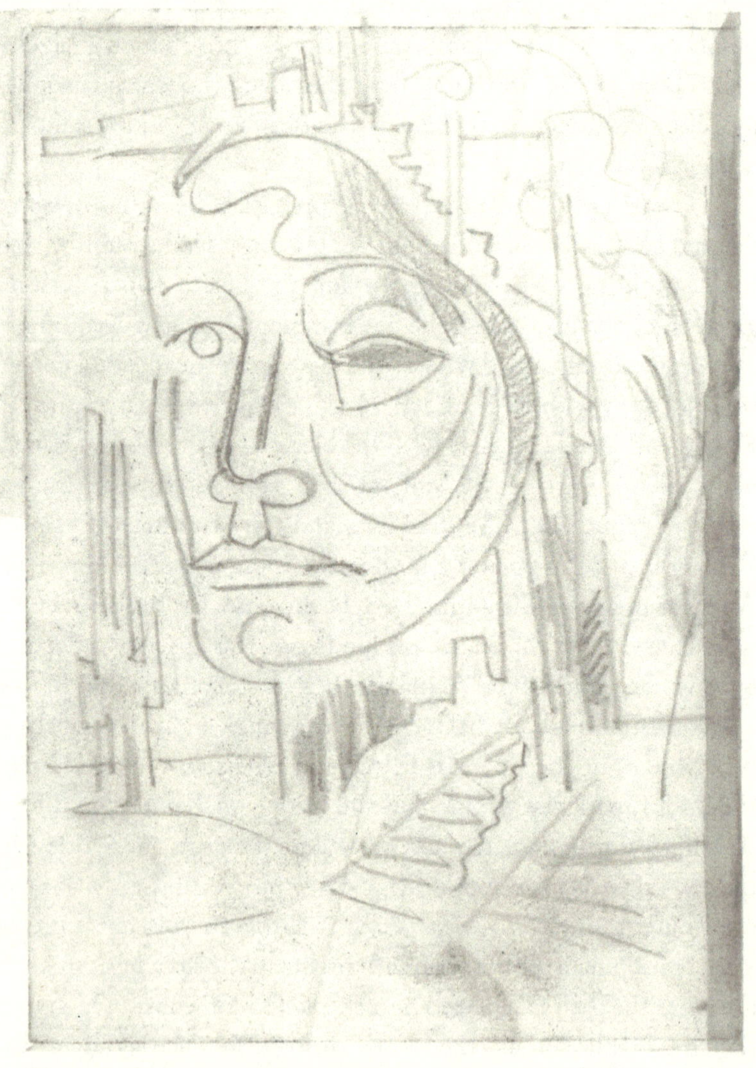

Figure 4b. Albert Flocon, preliminary drawing for the *Paysages* frontispiece (18 x 12.5 cm). Courtesy Catherine Ballestero.

the head, allows two more tiny workers to climb up into regions unseen, while a third leads down into the building's catacombs. In the deluxe edition, printed on China paper, this page of the series contains an additional miniature at the picture's lower left edge

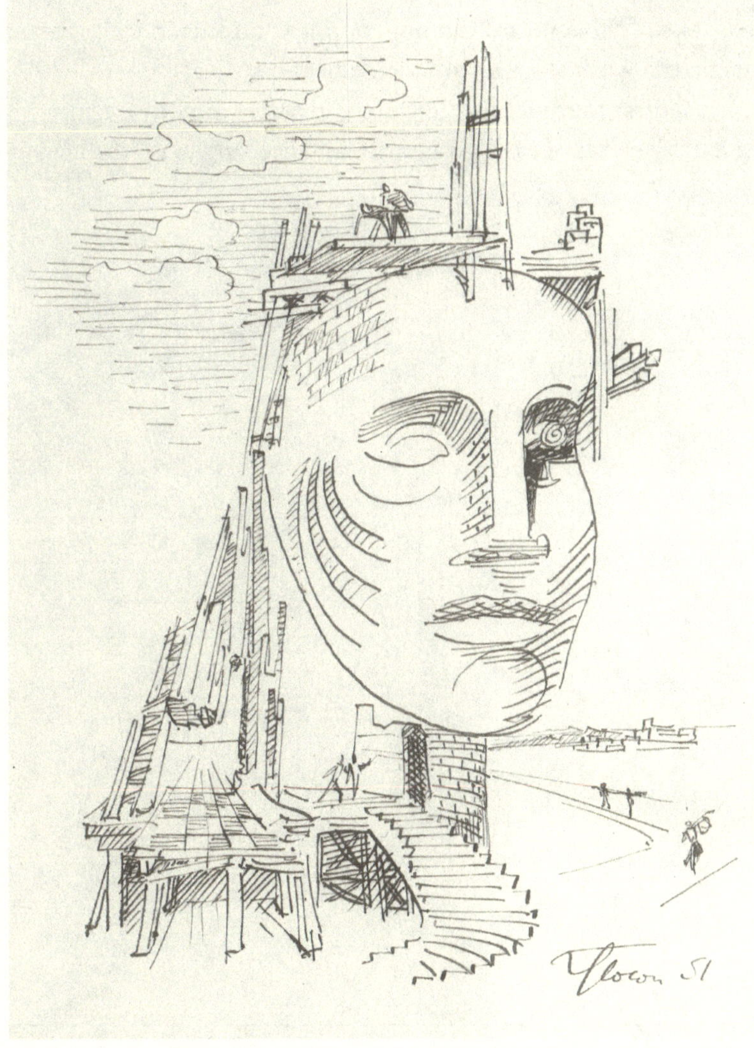

Figure 4c. Albert Flocon, drawing in ink for the *Paysages* frontispiece (20.5 x 15 cm), 1951, inserted into copy no. 7.

(fig. 4d) as a graphic annotation and counterpoint to the workers on the roof. The tiny figure points triumphantly to the unfinished colossus. The strangest aspect of the plate, however, is Minerva's eyes. They are open but empty—they see nothing. Flocon's Minerva embodied what he thought he had not yet achieved, "the wisdom that I so lacked."[26] Bachelard did not comment on this image. It stands outside the work as an enigmatic dedication.

The first four pictures in the main series directly engage with the four elements of earth, water, air, and fire, applying "the idea that

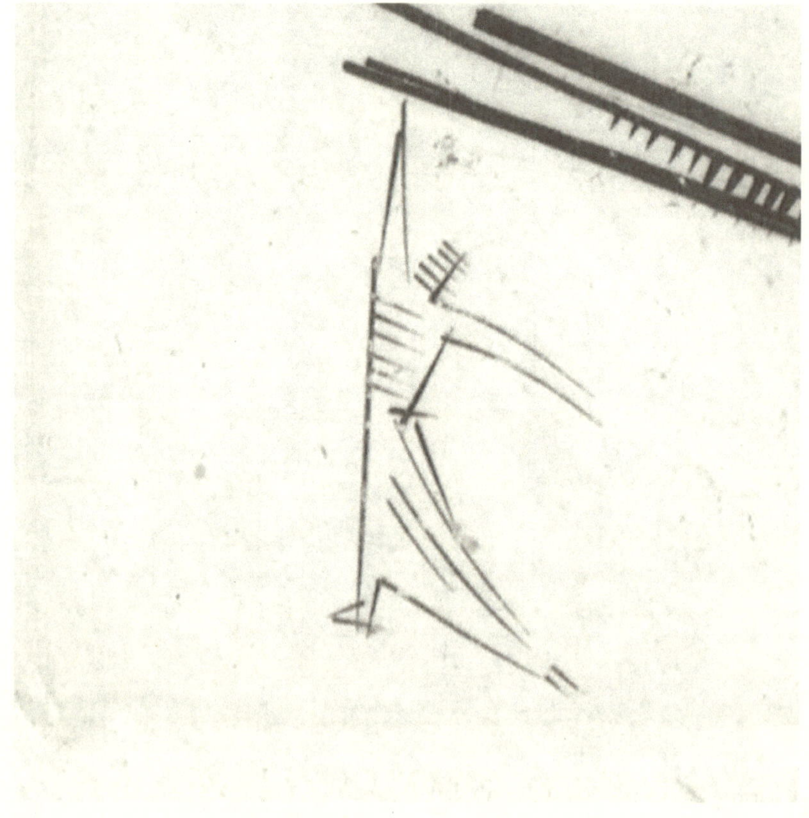

Figure 4d. Albert Flocon, frontispiece in the deluxe series of *Paysages* on China paper (detail, 2 x 1 cm), inserted with copy no. 7.

one thing can signify another."[27] The subsequent plates do the same in varying, more indirect forms. Flocon was asked during a radio interview on the book's appearance in May 1950 whether he had chosen the arrangement of the fifteen plates himself. He explained: "Gaston Bachelard was the one who thought up the order. I gave him an unsorted bundle of engravings, and he found the common thread to make a book of them."[28]

Let us look a little more closely at the fourth engraving, on the theme of fire (fig. 5a). Here, the scene is dominated by a dancing couple rising up from the ground in an eddy of motion. A flame turns into a whirl of smoke that scatters shards of fire and finally mingles with the body of the female dancer. It echoes the dynamic movements of the dancers while binding them together into a pirouetting loop that inscribes itself as a heart-shaped contour.

At first glance, we see just a single, conjoined body. Perhaps this was what disconcerted Bachelard; at any rate, the image gave rise to a series of letters during the summer of 1949, which Bachelard was spending as usual in Dijon.[29] In this correspondence, Bachelard reminded Flocon to let him have the "ink drawing *Fire*," which Flocon had omitted to give him before departing. Flocon accordingly sent an ink drawing, to which Bachelard responded immediately with a curt comment—"Your drawing remains confused"—and the request to see a "proper print" before beginning his reflections.[30] He pressed the point ten days later, on August 18, and repeated his impression that the drawing was "confused," though he thanked Flocon for confirming that he does not regard fire as "an animal." In matters like these, Bachelard remarked, "my long experience tells me that one must always rely on the print."[31] He was in a hurry to return to "my book," the text he was then working on.[32] Bachelard must have received what he had asked for in the days that followed, because he was able to send Flocon a copy of the completed manuscript on August 26. He mailed the manuscript as a sample to the publisher in Switzerland at the same time and hoped that Flocon would send "a quick word between two meditations of your treatise on the engraver."[33] Apparently, not only was Bachelard combining his reveries on Flocon's images with his work on a book about the philosophy of science and technology, but Flocon, while still finalizing *Paysages*, had already begun his next

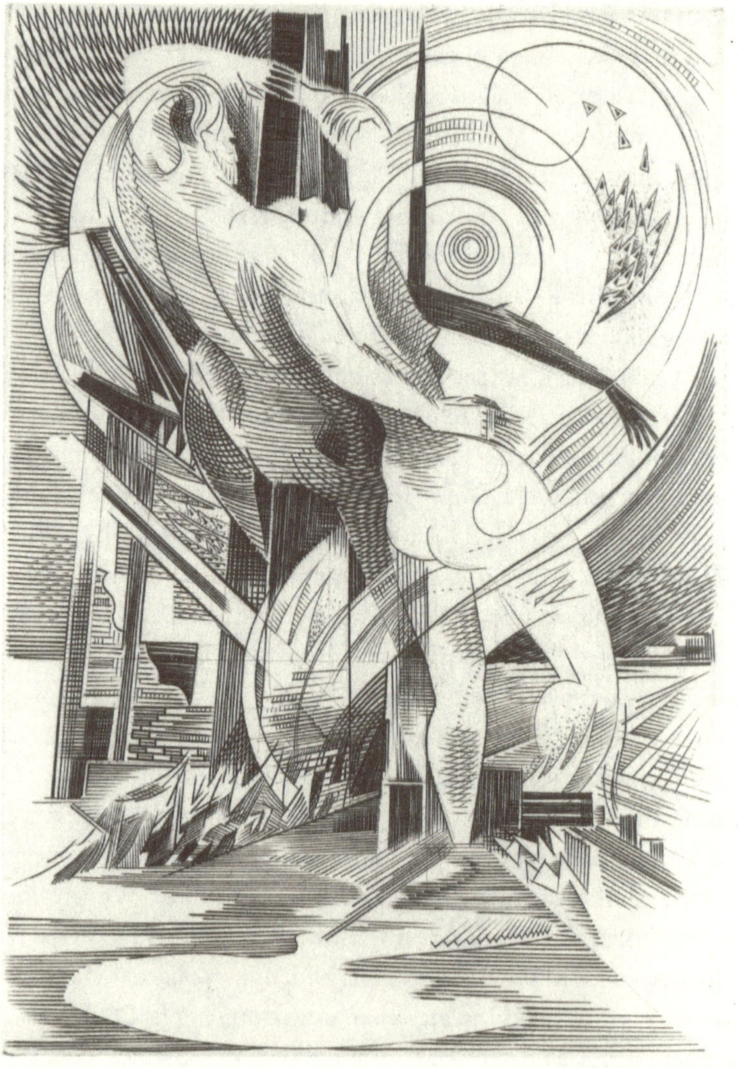

Figure 5a. Plate 4 (18 x 12.5 cm) in Bachelard and Flocon, *Paysages*, 39.

project, an essay on the activity of engraving. On October 2, Bachelard asked Flocon for three copies of the proofs when they were ready, to be read by himself, his daughter Suzanne, and a friend "who is very scrupulous on typographical details."[34]

It is not possible to know the exact state of the "Fire" image in summer 1949, when Bachelard was reflecting on it, but we do have a sheet of preliminary sketches made in 1948 (fig. 5b).[35] This fascinating document permits us to retrace in four steps the unfolding of the fundamental gesture upon which the completed image rests. That underlying motif is a flame ascending from fiery depths, its form progressively stabilized by vertical and diagonal elements, without a human figure yet having come into view.

The ink sketches were made on the back of a notebook page that is covered with undated penciled text (fig. 6). Whether this text should be read in tandem with the visual idea on the reverse of the sheet remains an open question. What is indisputable is it concerns Flocon's experience of the war. It is about the "responsibility of artists" in a situation that robs them of power and yet forces them to act. The crucial sentences run:

> This is not a question of principles, but a question resolved by facts. Hic Rhodos, hic salta. Hinc et nunc. There are situations in which it is absolutely impossible to equivocate. When the great catastrophes swoop down on us, events compel us to take sides. Then we see very well that nothing is absolutely gratuitous. Each of our previous actions, thoughts, and gestures has prepared the way for the conflict's conclusion. In all this machinery, our freedom is minimal. We are at once a motive power and caught in the gears; we can only give more or less brilliance to our action. In one sense, "everything is written"; in another, everything is to be done.

At the core of Bachelard's meditation on this plate is his remark that Flocon's fire does not conceal an animal, as is so often the case in mythologies of fire: "One striking feature of our engraver's cosmology is the absence of any representation of animals. Flocon proceeds straight from material forces to human forces."[36] The transition between material and human can be seen with great clarity in this image, in which the upward-coiling blaze blends into the right

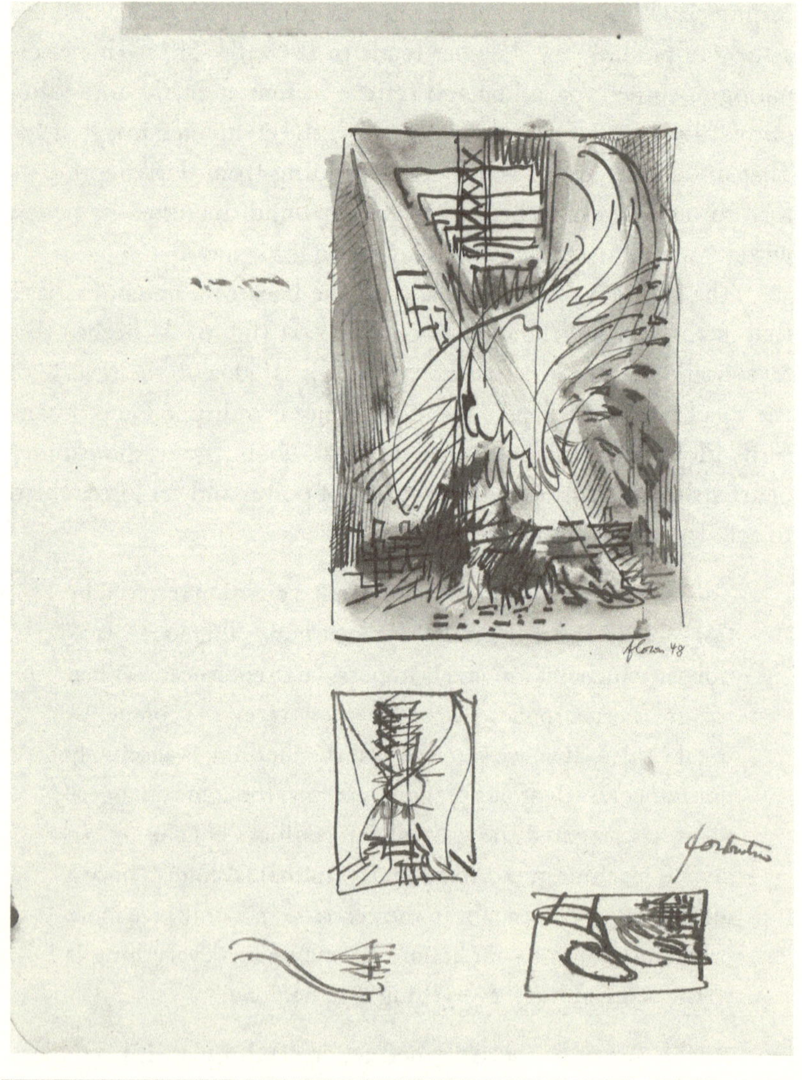

Figure 5b. Drawing (19.5 x 14.5 cm), inserted with copy no. 12 of *Paysages.*

arm of the dancing woman, which in turn is transformed into the left arm of the dancing man and thus into the counterflame formed by the bodies. The legs of the human figures, too, disappear into the flickering fire. The lopsided motion of the flame in the early sketch

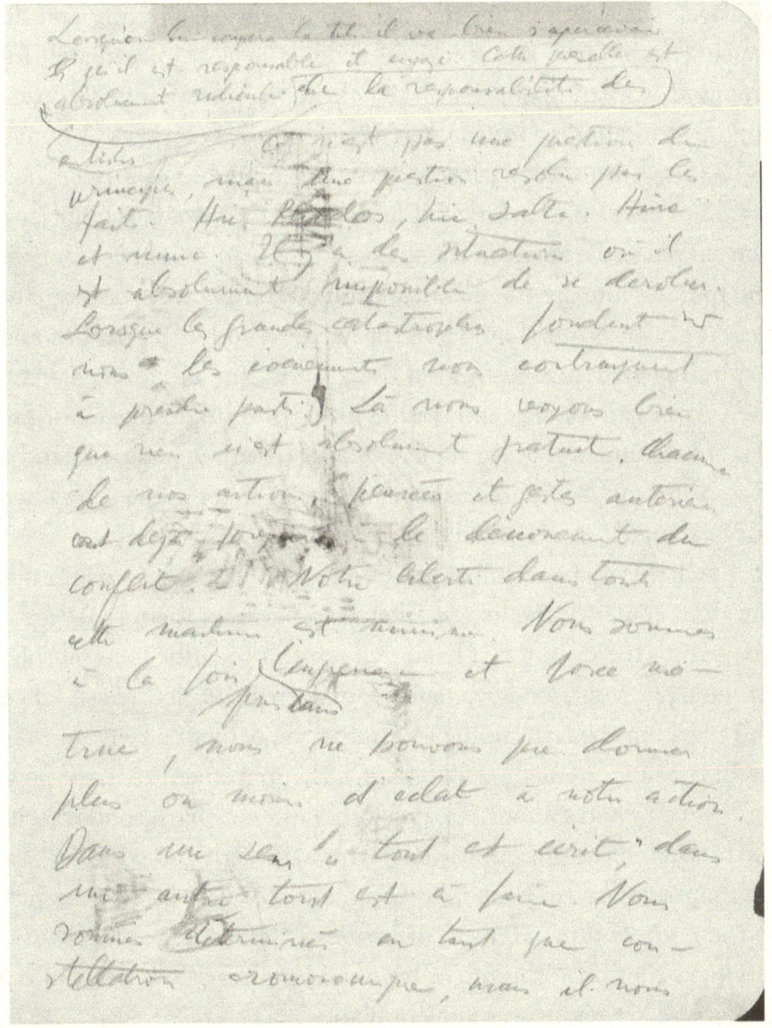

Figure 6. Notebook page with pencil script, undated, the reverse of the page of sketches reproduced in fig. 5b.

has now become symmetrical. "Love inflaming!" Admitting that the metaphor is "poor and hackneyed," Bachelard shows that it "takes on fresh life as soon as the artist *draws* it, as soon as, beneath his hand, it becomes truly *movement*. The couple becomes a vortex."[37]

Although no animals are to be found in Flocon's pictures for *Paysages*, there are plants and humans aplenty, encircling and devouring each other. This is exemplified by the plate immediately following the opening quartet on the elements, plate 5 (fig. 7a). Here, botanical and human forms intertwine. The arms of a brawny human torso grow into the gnarled and looping branches of a tree; those branches cast off five-fingered leaves that fly toward the horizon. The curved space between the torso and the arm-branches elongates and mutates into another hand, seemingly coming from nowhere, its fingers grasping one shoulder of the headless body. One leg becomes a spiraling structure resting on the ground, the other reaches into the earth as a root. The impression arises of a jungle-like engulfing and entanglement that, with increasing abstraction, iteratively and as if seen through a window, is repeated twice more at the spot where our gaze meets the horizon.

All these elemental ingredients grow out of the ground at the front center of the image. They pass between the torso's legs to merge into the deep forest before which this gigantic metamorphic struggle plays out, its movements dominating the picture as a whole and marking the left middle ground as the helix of a twisting tree. In all the ubiquitous transformations, the movement remains powerful yet soft and sinuous. This pliability offers a sharp contrast with the motion in the preceding plate, which, though also winding, was far more aggressive. Bachelard comments on plate 5: "We are here at the node of a metabolism of images."[38] He offers a deeply evolutionary dictum: "The life of root and bud lies at the heart of our being. We are really very ancient plants."[39]

In a similar process to the "Fire" plate, and confirming Bachelard's intuition, Flocon arrived at this picture from an initially rather undifferentiated study of motion (fig. 7b). Later in its development, it begins to take on a shape tending toward the outline of a struggling couple, the dynamics of its movement intimated by a broad S-shaped ribbon stretching across the page (fig. 7c). Only in the course of the engraving's realization do the

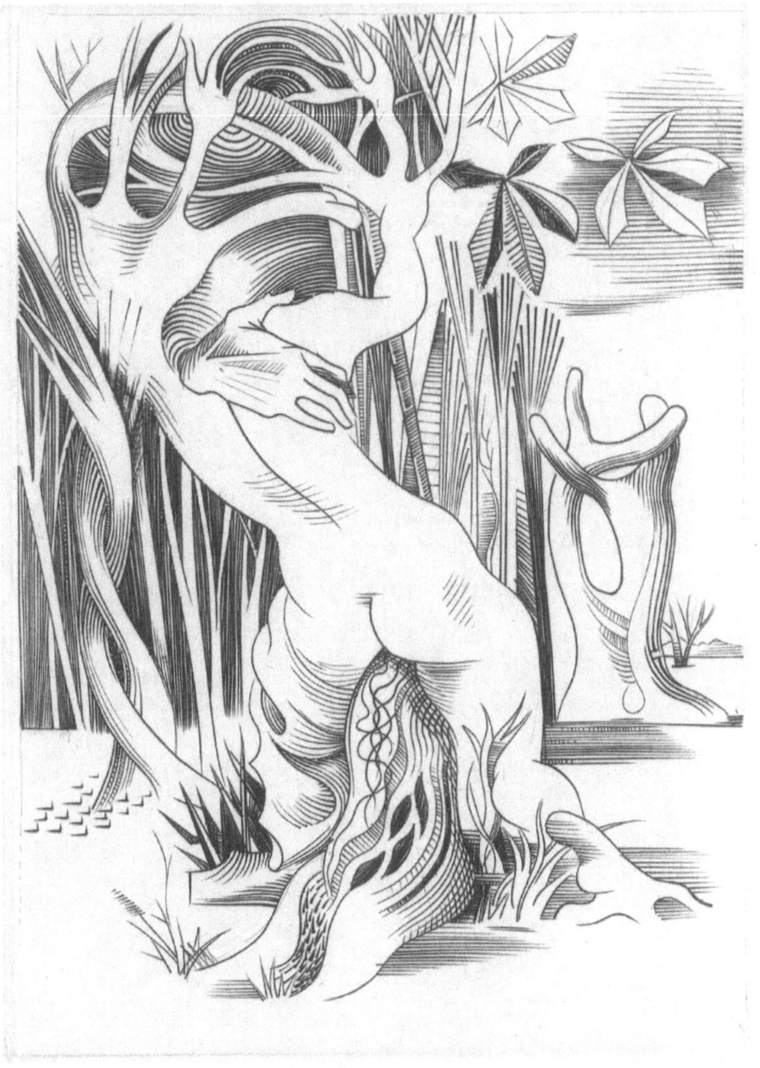

Figure 7a. Plate 5 (18 x 12.5 cm) in Bachelard and Flocon, *Paysages*, 45.

vegetable elements that so strikingly define the finished form attain their prominence, their parity with the body. In the radio interview already mentioned, Flocon summarized this process, the challenge of a form coming into being, in the following words: "There is a vague point of departure that might be an idea,

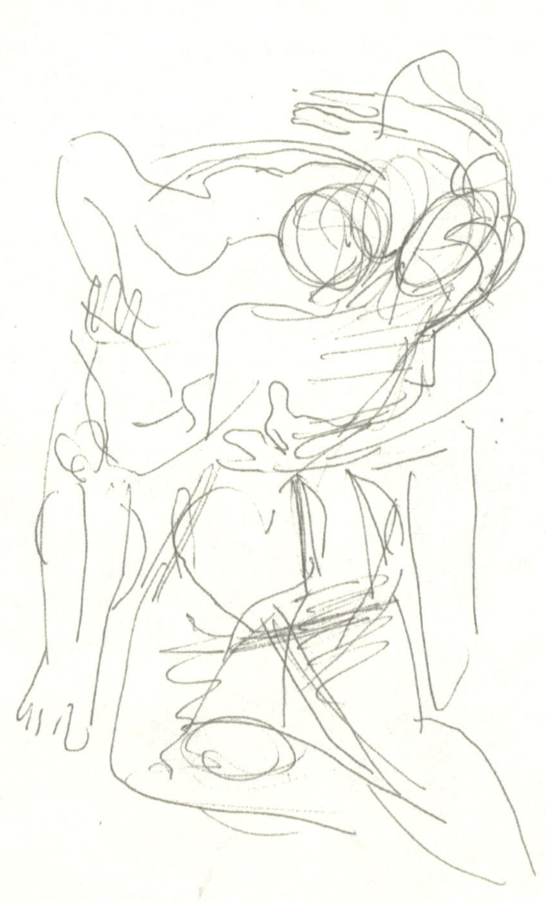

Figure 7b. Albert Flocon, preliminary sketch for plate 5 (27 x 20.5 cm). Courtesy Catherine Ballestero.

but is not yet formulated; something that takes shape as I trace it into the copper."[40] The dynamics of the hand are organized around a blind spot, what Gottfried Boehm has called a "kernel of blindness,"[41] just as language is organized around an "asemic kernel," in Rodolphe Gasché's trenchant phrase.[42]

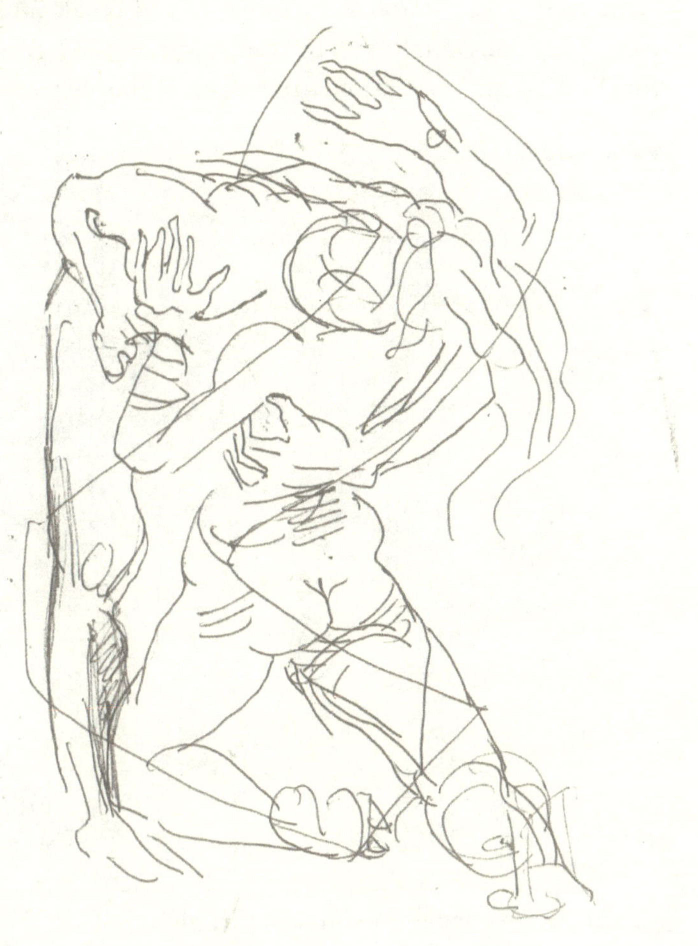

Figure 7c. Albert Flocon, preliminary sketch for plate 5 (27 x 20.5 cm). Courtesy Catherine Ballestero.

In this plate, Flocon may have been remembering Bachelard's chapter on "The Root" in *Earth and Reveries of Repose*. That chapter begins by evoking the image of the inverted tree, the "tree growing upside down."[43] Bachelard cites Fernand Lequenne, who himself cited Henri-Louis Duhamel du Monceau—the eighteenth century's great specialist on the Salicaceae—and his experiment with a year-old willow, turned upside down "so that the branches would become roots and the roots in the air would break into bud."[44] Flocon's engraving might also be read such that the figure's arms, jutting into the air, spring from a root system, and that leaves begin

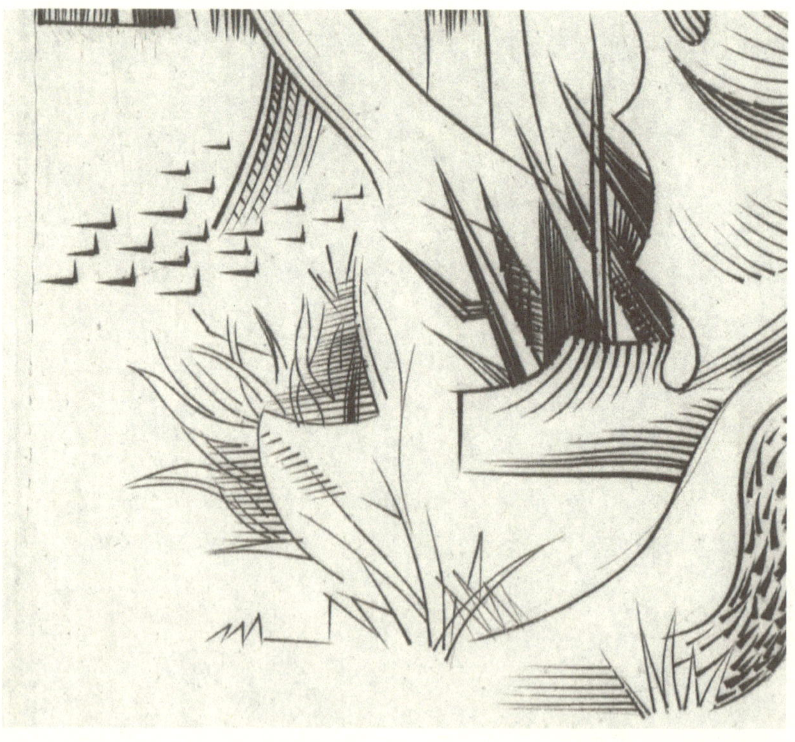

Figure 7d. Deluxe series on China paper, plate 7 (18 x 12.5 cm) in Bachelard and Flocon, *Paysages*, inserted with copy no. 7, detail (approx. 3 x 3 cm).

to sprout wherever the figure touches the ground—a doubly inverted metamorphosis.

In what sounds like an echo of this engraving, a letter from Bachelard in summer 1950 describes his work and garden in Dijon, where he is enjoying a "hard-working rest": "I am grappling with several projects without being able to choose. . . . I'm not even managing to master my garden. Everything is growing inopportunely—trees, weeds, thistles, brambles. I suppose the horticulture I propose is imaginary. But really, there are too many images around me. A miserable bindweed has suffocated a young peach tree, and everything is going the same way on this mad planet."[45]

But in landscape no. 5, it is not only the vegetal and human forms that continually transmute into each other. Forms are also continually and ungovernably growing out of interstices and gaining spatial contours; in turn, they themselves become interstices, dissolve. Metamorphosis rules the bodies, but also the actual space of the engraver and his scope to shape that space with the sheer element of line. Here we find a generalized continuation of the reflexivity that defined Flocon's engravings of hands: the grasp transforms not only the things that are grasped, but also the form of grasping itself.

Let me close by mentioning one detail. Like the head of Minerva, this plate also has a small graphic annotation, added to the deluxe series on China paper that Flocon enclosed with a selection of copies. In this case (fig. 7d), the comment is just one small extra tuft of grass in the lower left foreground. That tuft has plenty to say, however. Not only does it hint *in nuce* at the infinite play of metamorphic possibilities; it also shows how a few simple lines and marginal inscriptions can produce a new space within space. One almost sees yet another hand in the process of stretching out its fingers.

ON ENGRAVING

The craft of the copperplate engraver is the subject of *Traité du burin*, 1952,[1] in which Flocon undertook the experiment of analyzing in detail "what happens at once in the head, the hands, and the eyes of an engraver."[2] The treatise thus examined his own work and, "more precisely," offered "a continuation of Bachelard's manifesto 'À la gloire de la main.' "[3] Flocon had been working on the project for quite some time, and had published a short "Éloge du burin" two years earlier in *Art d'aujourd'hui*. That outline of the study magisterially asserted that in art as a whole, and most particularly in copperplate engraving, "the conceptual struggle is inseparable from the struggle with matter."[4] The seven chapters of the full *Traité du burin* address "Engraving" as a primordial gesture, "Sharpening the Burin," "Mirroring," "The Proof," "Drawing the Stroke," "Meaning and Form," and finally "In-Formation." It presents the case for Flocon's claim that the artist's freedom consists in overcoming impediments, just as Bachelard saw the quintessence of the scientist's work, thus of the "formation of the scientific mind," as being the ability to overcome "epistemological obstacles."[5] Flocon concludes: "The future artist will have need of obstacles: from those that are posed by the material being worked to those in the past, present, and future that society, the world, induces in the mind of the worker himself."[6] This is unmistakably Flocon's response to the state, as he saw it, of art in his epoch, which he had earlier castigated as "an era of debris, of the charm of matter in dissolution, of the refinement of trimmings."[7] Flocon remained true to his conviction. Thirty years later, he reiterated: "Why the choice of this thankless tool of all things, this burin—so unyielding, hard, aggressive, resistant, intractable? That's just it! In times of laxity . . . a hard craft is a good guide. Wanting to be free does not make you free."[8]

Printed by Frazier-Soye and hand-pressed in Georges Leblanc's workshop, the *Traité* was published under the supervision of Georges Blaizot at the distinguished Paris company established by Auguste Blaizot. Flocon explained that his book was "created not so much to teach copperplate engraving as to enlighten the author himself about that work," being a "transcript of his explorations in a material and a technique and his research on a flexible and open method."[9]

The *Traité* consists of copperplates detailing the various steps in the engraver's work, accompanied by texts written by Flocon himself. It marks the beginning of his activities as an artist of the written word. Flocon had little interest in the relationship between word and image in the traditional sense, whether as descriptions of his own pictures or as illustrations of his own text. Rather, he found that the text took on its own logic, one that paralleled the act of engraving: "I soon realized that writing is to prattling as engraving is to doodling."[10]

Among the *Traité*'s plates, we find the hand of the engraver in a new set of variations. Flocon referred to one as "trihedron" (fig. 8), the other as "plow" (fig. 9).[11] These are paradigmatic incarnations of the two aspects of engraving that had caught Bachelard's attention: the motif of constructive emergence on the part of the worker, and the motif of resistance on the part of the matter being worked. The first plate embodies construction. One might say that it "handles"—to go back to the primary meaning of the word—the three dimensions of space, in the form of a mighty hand that is part of an even larger building site. The little figures in the site's interior seem like marionettes controlled by the fingers. Unlike the hand that concluded *Perspectives*, this one has detached itself completely from the body and become integrated into the construction site. The human bodies are now mere appendages.

The second plate articulates the resistance of matter. With willful and emphatic austerity, it shows the effort of the engraver's hand guiding the burin, defied by the blank copper. The flat surface folds, so to speak, into furrows under the plow. It is from those furrows that everything must grow. The shaving driven out of the

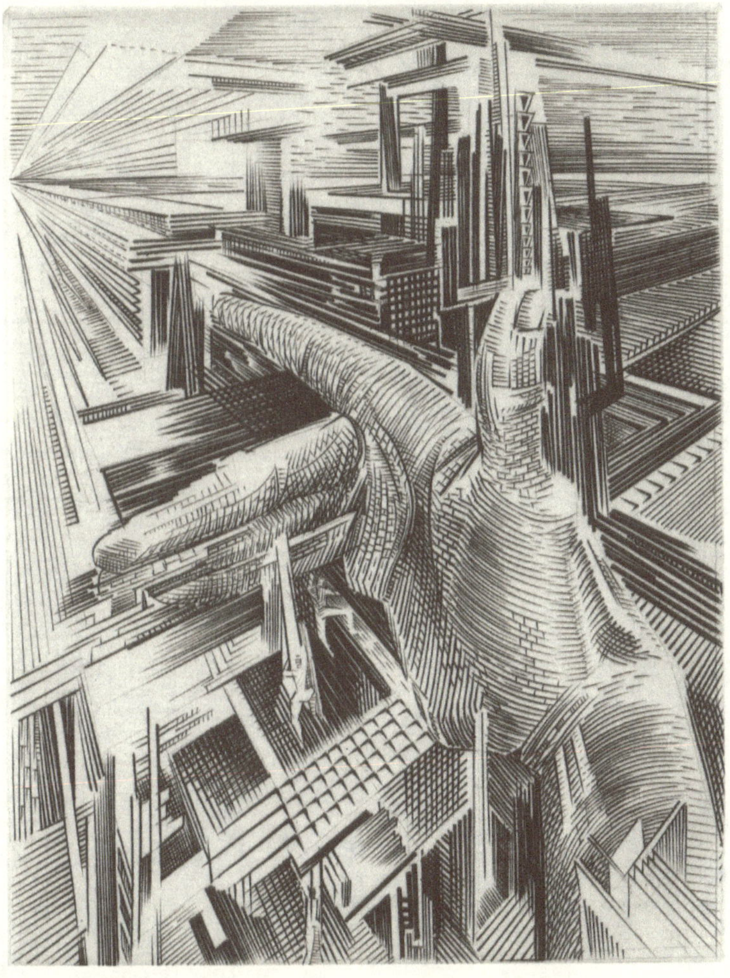

Figure 8. Plate "Le trièdre" (12.5 x 9.5 cm) in Albert Flocon, *Traité du burin. Avec une préface de Gaston Bachelard* (Paris: Librairie Auguste Blaizot, 1952), 21, copy no. 16 (on Vergé de Montval) of 260.

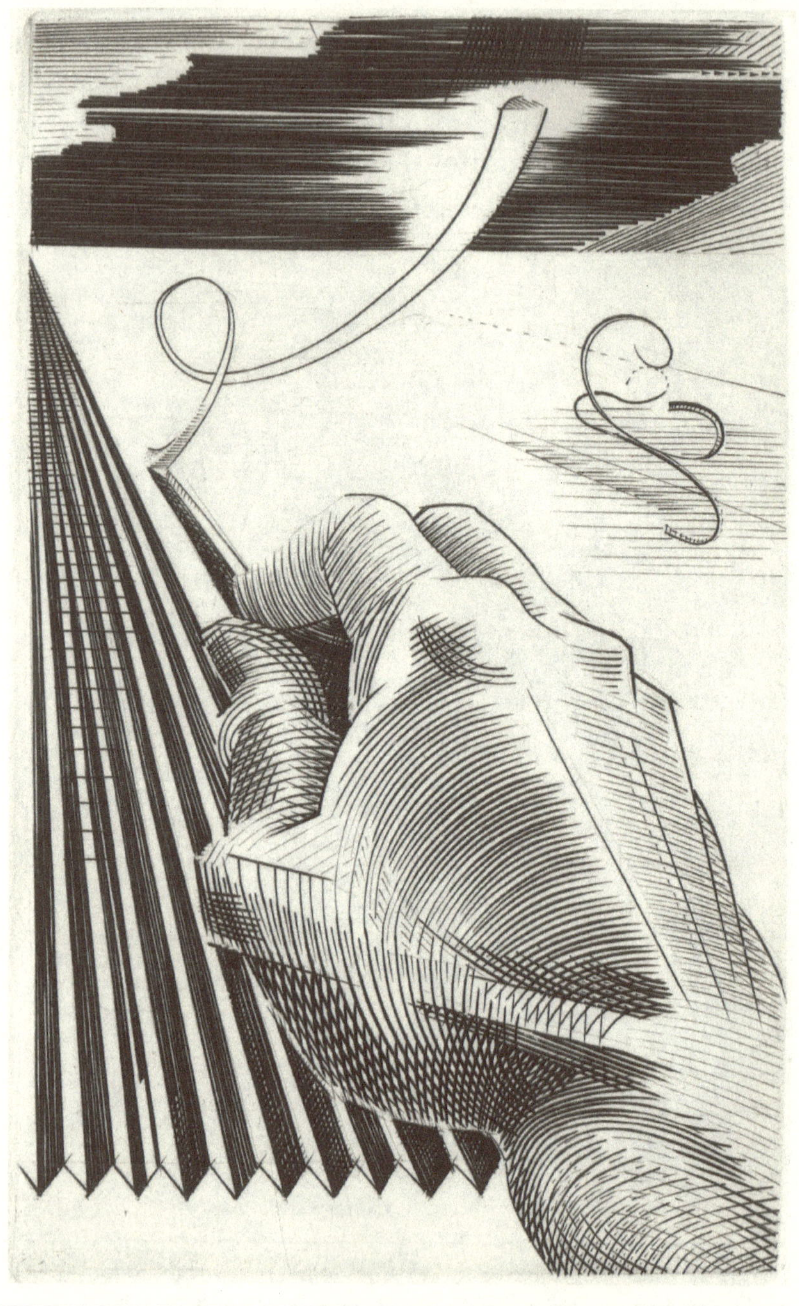

Figure 9. Plate "La charrue" (13 x 8 cm) in Flocon, *Traité du burin*, 27.

plate by the tool curls across the horizon into the dark sky, illuminating it brilliantly like summer lightning. Interestingly, it is the waste material from the labor of engraving that, through the dance of shavings, breathes motion into the picture. The metal shaving both evokes the incisiveness of the procedure and allows us to grasp why Flocon called copper engraving "an extremely abstract technique."[12]

But the hand does more than construct, more than only score and incise. As the next plate shows, it also "caresses" (fig. 10).[13] Here, the counterpressure of matter, the bark of a gnarled branch, has to be sensed by probing fingertips. The movement continues right into the crooks of the fingers. That is the other side of the confrontation: resistance as empathy, involvement. "Of course, there is at first a hesitation, a responsibility that one suddenly senses when cutting into something as beautiful as a shining, well-polished copper."[14] The closing lines of Hedwig Wingler's contribution on the 1992 Flocon retrospective in Metz seem perfectly attuned to these three hands: "The 'pleasure in becoming' of the material encounters the 'pleasure in producing' of the hand, which is an eye that can measure, plow, love."[15]

Bachelard contributed a preface to Flocon's *Traité*, writing on what, in an inspired phrase, he calls "digital will" or the "will of the finger" (*la volonté digitale*). Exposed to the resistance of "matter that is real and strong," Bachelard remarks, the engraver "can never be passive; he copies nothing; everything must come from him, must be produced by him with a minimum of lines, surfaces he has to create by outlining them; to bring out his solids he has only the superposition of perspectives."[16] The engraver is "inspired by the nascent form";[17] conversely, he calls it into being. Flocon commented that such work was the result of "relentless effort, material and mental."[18] The short but far from modest "Éloge du burin," which can be regarded as preparatory work for the *Traité*, announces boldly that copperplate engraving, "with its sole and singular element—the line—and its technical constraints, could open up a route along the whole length of which matter and spirit, in their dialectical

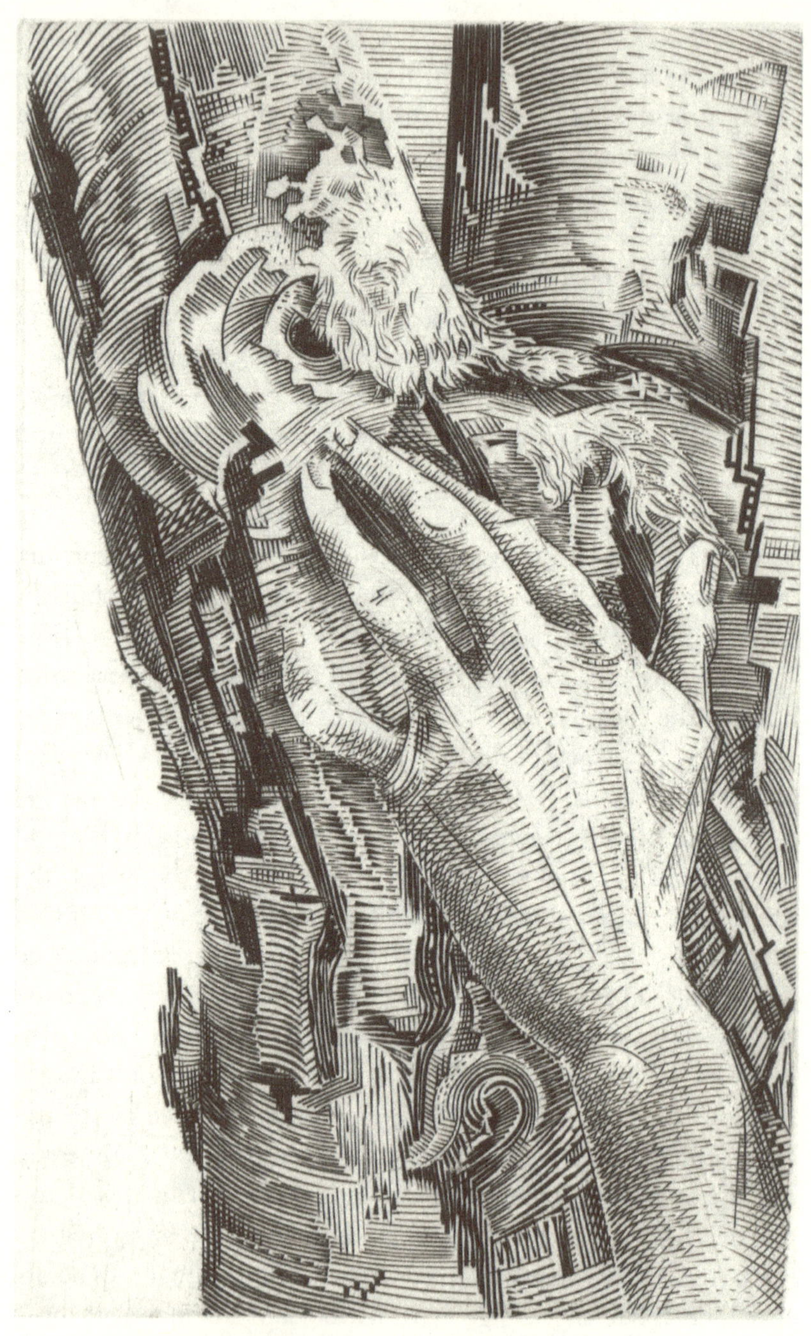

Figure 10. Plate "La caresse" (13 x 8 cm) in Flocon, *Traité du burin*, 35.

interdependence, would be able to give art a better grip on the world and a new raison d'être."[19]

Bachelard had devoted the second chapter of *Earth and Reveries of Will* to a series of reflections on the character of the tool and its challenges. "Tools awaken in us a need to act *against* something hard."[20] There, he drew on the work of archaeologist André Leroi-Gourhan, whose *L'homme et la matière* proposed a classification of tools according to the hardness of the materials they worked—because "the material conditions every technique"[21]—and on the fruits of his literary reading, which ranged from Georges Buffon, to Victor Hugo, to Éluard and Tzara. Flocon's *Traité du burin* now offered him the opportunity to share in an "awareness of the implement" (*conscience d'outil*) by tracking the verbal and visual labors of an artist.[22] This terse formulation, its suggestive power fueled by a grammatical play with the genitive case, compresses the whole problematics of the tool. The relationship between knowledge and hand, between body and mind, is a reciprocal one; at the same time, the issue is no longer simply knowledge-making by the hand, but knowledge-making by its technical extension. For the attentive reader, the short and almost staccato sentences of Bachelard's introduction to the treatise seem to mimic the scratches that he saw the tool scoring so energetically into the metal.

Flocon, in turn, admitted that Bachelard's work had inspired him to write the *Traité du burin*. "I was very impressed by Bachelard's writings," he reflected in his contribution to the 1970 Bachelard colloquium held in Cerisy-la-Salle, "and I put together a little 'engraver's treatise,' in which I wanted to convey an emotion and a pleasure that arose not so much from contemplating the finished work as from implementing it."[23]

Focillon's "In Praise of Hands" picks up on the idea of the artist, amid the technical elaboration of our world, prolonging the archaic gesture of ancient crafts and keeping them contemporary. Flocon, too, starts from the observation that for artists, "tools have barely changed over thousands of years."[24] He categorically repudiates all "nostalgic dreams," however.[25] For Flocon, the crucial point is not to keep a fundamental human experience alive through a material gesture, but his conviction that the "imaginative faculty"—the capacity to bring forth something new—"is very intimately connected to experiences of the hand" and can become manifest in ever new variations.[26] It is not, then, a matter of preserving something past, but a matter of the future: the inexhaustibility of an elemental gesture of knowledge. Flocon described the quintessence of his treatise thus: "It has become clear that art may be a special form of knowledge to a greater degree than is generally believed, and that its realizations are the outcomes of relentless effort, material and mental."[27]

POETICS

In the 1942 book *Water and Dreams*, subtitled "An Essay on the Imagination of Matter," Bachelard distinguishes between "a *formal imagination* and a *material imagination*." The distinction is a genetic one, based on the sources of the images concerned. Whereas the formal imagination is internal, Bachelard calls the images of the material imagination "images that stem *directly from matter*," adding that "the eye assigns them names, but only the hand truly knows them."[1] Joseph Noiret, a cofounder of the international artist group Cobra, stresses the influence that the "happy materialism" of Bachelard's books on fire, water, earth, and air exerted on the members of the group: "This happy materialism, which explains a number of works created by Cobra, imposes the idea that the *means* being used and the tracing *hand* take an immediate part in the inspiration of the creator. . . . This conviction explains Cobra's adoption of the very diverse material supports of writing and painting that play their role in creation. It is never with indifference that a creator chooses one type of material rather than another."[2] Applied to poetics, this means that the work of the writer is located in the same order as that of the painter and the sculptor—or of the artisan. The problem of the "relationship between material and formal causality" confronts the poet as much as the visual artist, for "poetic images also have their matter."[3] And matter is what must be explored: "In the depths of matter there grows an obscure vegetation; black flowers bloom in matter's darkness. They already possess a velvety touch, a formula for perfume."[4] For poetry, Bachelard concludes that "if a reverie is to be pursued with the constancy a written work requires, to be more than simply a way of filling in time, it must discover its *matter.* A material element must provide its own substance, its particular rules and poetics."[5]

Bachelard frequently returned to this distinction within the imagination. In his essay on *Earth and Reveries of Will*, we find "the imagination in matter" contrasted with that "in word" (*l'imagination parlée*), the "creative" with the merely "reiterative imagination," the "material imagination" with the "dynamic imagination."[6] The point is always to privilege the materialized imagination over its too-flimsy alternatives, to grasp it as the real, ultimately to represent it as labor. What is particular about the book on earth is that it represents the poetic labor itself, in the images of labor as they engage with the terrestrial element. The book is thus concerned with labor articulated, with the labor of the word (*le travail parlé*).[7] Here we find the special oscillating quality of this text, its ambiguity: the experience of "muscular pleasure,"[8] delegated to the implement of material labor—"when my knife likes to carve"[9]—recurs as a pleasure in images and its delegation to the linguistic acts of the poet.

For Bachelard, the decisive aspect of the writer's and the artist's work—and, incidentally, also the work of the scientist—is his opening up to the way ahead, incited by the material he has chosen to confront. To remain in the imaginary of matter and hand: it is the act of seizing. Bachelard's book on literary images of the resistance of terrestrial matter, of the "world of metal and stone, wood and rubber," distills this principle into an imperative for the art of writing: "Every image which flows from a writer's pen should contribute something new."[10] But the process of creating new images is not left to imagination alone; it is tied to the material process of writing itself and is thus "earthed" in its turn.

In his third book on the elements, *Air and Dreams*, where the elements are called "the hormones of the imagination,"[11] Bachelard uses powerful imagery to underline the same connection. Both writing musicians and "silent poets," he argues, "hear what they write at the same time as they are writing it, in the slow cadence of written language." They can trust that "their pen would stop of its own accord if it encountered a hiatus, that it would refuse to write unnecessary alliterations."[12] The resonances with the surrealist theme of *écriture automatique* are loud and clear in sentences like this one on the experience of written reverie: "What he had intended to say is so quickly replaced by what he discovers himself writing that he feels clearly that the written language is creating its own universe."[13] Imagination needs to be led by the writing implement in order to surpass itself, to gain possession of its own emergence. The field of possibilities thus unfurled serves an "epilogism" that finally—in one of the arresting expressions that seem to escape the philosopher as he writes—drives Bachelard to exclaim: "The pen sings!"[14]

In his late work *The Poetics of Reverie*, Bachelard espouses this figure of surpassing—the aspect of what is unforethinkable but very far from unwritable—even more strongly than he did in the references scattered through his early poetological writings on the four elements: "In writing, the author has already performed a transposition. He would not *say* what he has written. He has entered . . . the realm of the written psychism."[15] A written word, says Bachelard in the same work, "is a bud attempting to become a twig. How can one not dream while writing?" However, he immediately adds laconically, alluding to the philosophy of the tool described earlier: "It is the pen which dreams."[16]

From the very outset and in quite some detail, Bachelard argued in his epistemological writings that instruments play a rather similar role in the development of knowledge. Despite this parallel, however, he always maintained that the two creative worlds, the world of the sciences and that of the arts, must be treated separately (and, of course, he followed exactly that division in his own oeuvre): "Two vocabularies should be organized to study knowledge and poetry. But these vocabularies do not correspond."[17] This seems unambiguous—yet the negation of correspondence does not mean a negation of complementarity. Well versed in the physics of his era and the meaning of its notion of complementarity, Bachelard hoped, starting from his very first work on poetics, "to make poetry and science complementary, to unite them as two well-defined opposites."[18] In an interview from the early 1950s, he put it as follows: "For me, if art is one of the sciences, science is an art in its own right."[19] On this point, Jean-Jacques Wunenburger has recently commented: "Bachelard's philosophical project is to exemplify through two limit situations, scientific practice and oneiric experience, the capacity of our representations to lead us constantly toward the transformation of our knowledge and convictions, and thus the transformation of our existence."[20]

At an early stage in his epistemological oeuvre, Bachelard began to move away from an idiosyncratic psychoanalysis toward an equally idiosyncratic phenomenology. One might describe it as a phenomenology of matter, a phenomenology that, at the suture connecting the self with the world, does not privilege intentionality over matter but, on the contrary, allows itself to be guided and beguiled by matter's own elemental constitution. In his later work, Bachelard completely abandoned psychoanalysis and its reading of symbols or symptoms. Rather than as a carrier of "psychemes," literary expression came to be regarded by Bachelard primarily in its own intractability—in what François Dagognet once acutely described as a "perpetual progression beyond itself."[21] In Bachelard's own words: "Literary life in particular is all adornment, ostentation, exuberance. It develops endlessly in the realm of metaphor."[22] Introducing Bachelard's letters to the poet Louis Guillaume, Jean-Luc Pouliquen remarks that Bachelard had a heightened sensitivity to this expansiveness of language, and even in his old age retained "the capacity to marvel at new expressions."[23]

The germ of a transition from a more psychoanalytical to a more phenomenological register is already present in Bachelard's literary study on the imagination of earth, written in the middle of his career. The shift is only fully accomplished, however, in the late *Poetics of Reverie*. There, the concept of dreaming or reverie takes on a pivotal role. It designates the experience of allowing oneself to be led along the edges of poetic newness, a journey that brings the literary image into the world—or should we say fetches it out of the world? Georges Canguilhem offered an illuminating comparison: just as "error is not a weakness but a strength" of Bachelard's epistemology, in the same way "reverie is not smoke but fire" in Bachelard's poetics.[24] The meanderings of science and the daydreaming of poetry meet at a functional tangent that reveals the common motive force behind Bachelard's work in poetics and his work in epistemology.

At this point, we may be approaching what Canguilhem cautiously termed the "secret of the equilibrium" between Bachelard's "two predilections" and perhaps come one step closer to "the investigation of its still-hidden foundation."[25] Even if their "vocabularies do not correspond," as Bachelard himself stressed, there are nonetheless resonances and recollections between these two transgressive ways of relating to the world, which converge in one fundamental trait: the mediatized, material character of their provocation and of the challenges they engender. In an offprint of his preface to the French translation of Roland Kuhn's *Über Maskendeutungen im Rorschachschen Versuch* (On Interpretations of Masks in the Rorschach Test),[26] Bachelard wrote the following dedication on June 29, 1957—probably one day after receiving the collection *Hommage à Gaston Bachelard*, the first essay of which contains Canguilhem's words cited earlier: "To my friend Canguilhem, to convince him that in a well-balanced 'duplicity' of the side of the 'subject' and the side of the 'object,' phenomenology is a science of madmen and geniuses."[27] Bachelard is alluding to the gist of his preface, which he summarizes as follows: "Ambivalences are never simply juxtaposed. Between their poles a process of conversion of values is continually going on."[28] Incidentally, Flocon had met and come to admire Canguilhem during Bachelard's soirees at rue de la Montagne Sainte-Geneviève, as he notes in his memoirs.[29] For the visual arts, Flocon too was convinced that one cannot simply wait for inspiration to strike: inspiration "will arise as one moves forward, as one acts," just as "no theoretical explanation can replace practical experimentation."[30]

CASTLES IN SPAIN

"I have made another book with Bachelard," we read in Flocon's autobiography, "*Châteaux en Espagne*, for which he has written the commentary. These are the greatest copperplates I have ever made. I wanted to create a book of engravings, some of them bled off the page, flush-mounted like a photography book."[1] Paradoxically enough, the playful title, evoking the daydreaming "castles in the air," is what gives this bundle of sheets its programmatic coherence. The individual pages might also be regarded as a theatrical sequence: as a succession of acts revolving around formations that are always about to slip away over the horizon. They articulate proximity and distance at once. Like *Paysages*, the album was printed by Fequet and Baudier; Flocon stamped the plates himself on Georges Leblanc's hand presses. It was produced for the Cercle Grolier—"The Friends of the Modern Book," named after the sixteenth-century bibliophile Jean Grolier—and its president Paul Banzet. The invitation to the book presentation features a small double portrait of the engraver and the philosopher in one corner of the plate (fig. 11): Flocon in shirt sleeves, his hands on his sketchpad and his eyes fixed on an object out of sight; Bachelard with his arms crossed, gazing at once inward and out into the distance. There they stand, side by side, not deep in conversation but each immersed in his own vocation.

Describing the period when the book was taking shape (it appeared in November 1957), Bachelard exclaimed: "The stories I've been telling myself all winter as Albert Flocon brought me week by week the individual sheets of his album!"[2] The winter in question

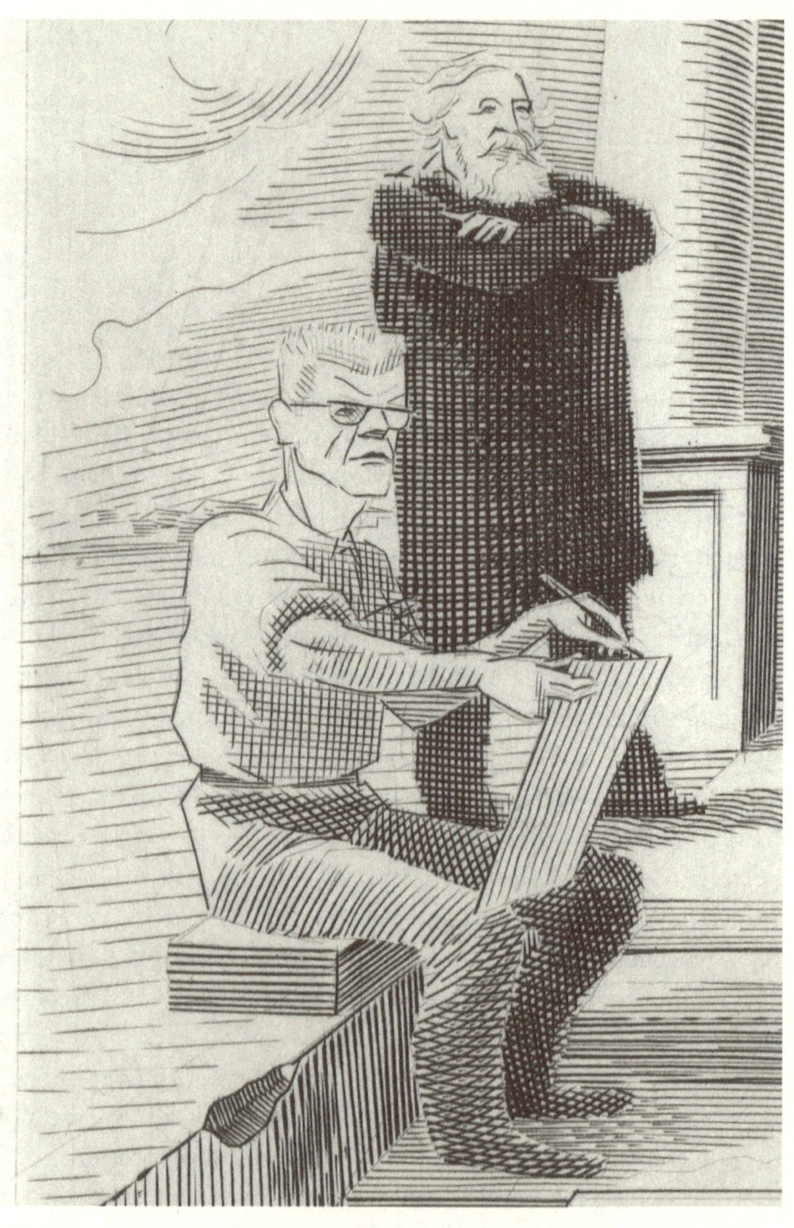

Figure 11. "Grolierii Amicorum," printed invitation to a banquet on February 15, 1958, detail of frontispiece (9 x 6.5 cm).

appears to have been 1955–1956, because in fall 1956 Bachelard wrote Flocon from Dijon to say that he had now finished the last missing page for the "Castles in Spain."[3] The book's printing appears to have been a slow business, and Flocon had already spent four years working on the large copperplates.[4] Bachelard's preface once again stresses what one might call the two men's unspoken cooperation—each working on his own account rather than in a close interaction between two ways of engaging with a theme. Bachelard tells the stories to himself. "Flocon, of course, never told me what his intention had been. He made me no speeches."[5] But Bachelard captures the substance of Flocon's enterprise very precisely when he confesses: "I like engraving for its own sake, autonomous engraving, engraving which is primarily not illustration, the kind I call in my philosophical ruminations autoeidetic engraving. For me this is the ideal form of the story without words, the distilled story. And it is because the engraving 'tells' nothing that it obliges you, the musing spectator, to do the talking."[6] The same gesture defines his own writing. In that sense, the subtitle of Bachelard's text for *Châteaux*, "The Philosophy of an Engraver," is well chosen. Here, the engraver becomes the philosopher of his own production, while the philosopher creates a twin world of sharply chiseled verbal images.

Flocon once wrote: "I have always had a fondness for building sites, works under construction; my whole life long I have built castles in Spain."[7] As Bachelard notes: "Flocon calls his collection *Castles in Spain*. He is inviting us to measure the distance between that which is seen and that which is dreamed; he is inviting us to move through what might be called *project-space*, to live in the *space-time of the project*. This is how, impenitent philosopher that I am, I would sum up Flocon's vision: Flocon is the engraver of the *space-time of the project*."[8] To project is to construct the construction, to attend to the act of building itself. Flocon "loves to capture an instant of construction when the construction has not yet been completed."[9]

These words are strongly reminiscent of a passage from *The New Scientific Spirit*, the key text of Bachelard's early period of epistemological endeavor, in which he characterizes the modern sciences as follows: "Above the *subject* and beyond the *object*, modern science is based on the *project*. In scientific thought the subject's meditation upon the object always takes the form of a project."[10] The point is elaborated a few pages further on: "A truly scientific phenomenology is therefore essentially a phenomenotechnology [*phénoménotechnique*]. Its purpose is to amplify what is revealed beyond appearance. It takes its instruction from construction."[11] Here we approach the essence of the sympathy between Bachelard the philosopher of science and Flocon the engraver. Bachelard regards Flocon's oeuvre as embodying an empowerment whose gesture he sees at work in the modern experimental sciences—and that, *mutatis mutandis*, also typifies his own philosophy of science.

It is, then, by no means the archaic aspect of the engraver's gesture that appeals to Bachelard, but rather its knowing seizure of the material and imaginary world. One might say that for Bachelard, reflection on copper engraving is something like a source of light in which his epistemology and his poetics can illuminate each other. In the domain of mark-making, labor and matter are present even more concretely than in the process of writing. This opens up a new space for the play of abstraction and concretion, in which scientific and poetic production may mirror themselves and each other. The motion involved is one comparable to the stretto of a fugue. Abstraction and concretion can be thought of here in the literal sense of taking away and bringing together, accumulating. Concretion is achieved by means of abstraction—and that, incidentally, is the basis of experimentation. This is overlooked when, as so often occurs, the abstract is banished to the banks of theory and the concrete to those of the experiment. Fundamentally, the material activity of experimenting consists in an effort of omission, out of which increment ultimately arises. Specification tips over into surprise.

In a kind of complement to the self-portrait as Minerva that preceded *Paysages*, Flocon prefixes *Châteaux en Espagne* with a portrait of Bachelard (fig. 12a). Most remarkable in this frontispiece are, again, the eyes. Slightly narrowed, they give the impression

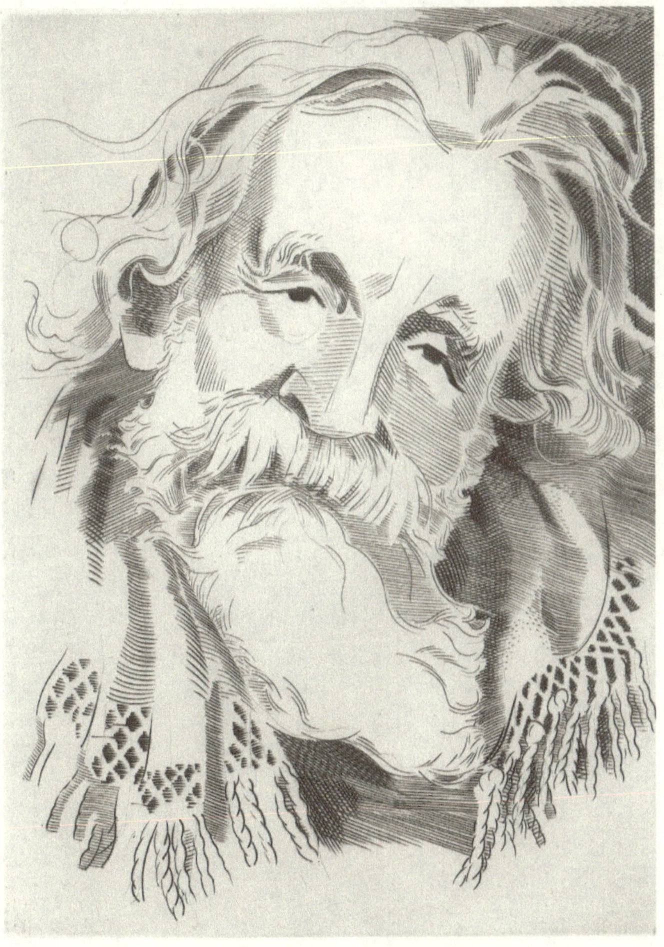

Figure 12a. "L'auteur," plate 2, frontispiece (19.5 x 14 cm), in Gaston Bachelard and Albert Flocon, *Châteaux en Espagne* (Paris: Cercle Grolier, Les amis du livre moderne, 1957), 8, copy no. 72 (printed for M. Henri Jaudon on Rives) of 200.

of looking inward and outward at the same time, as if connecting Gaston Bachelard's two worlds—in stark contrast to the blind eyes of Minerva in *Paysages* (fig. 4a). Two preliminary studies for the Bachelard portrait (fig. 12b, fig. 12c) bring to light a further

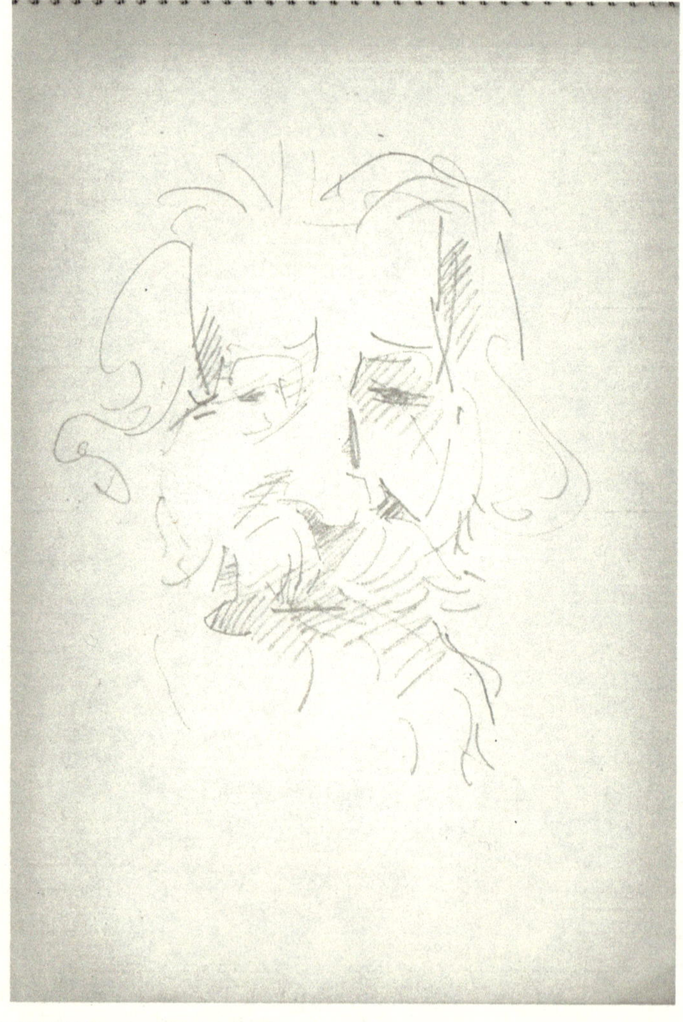

Figure 12b. Preliminary study for the portrait of Gaston Bachelard (23.5 x 15.5 cm). Courtesy Catherine Ballestero.

noteworthy aspect of the image. Flocon has played with the tilt of the head, gradually shifting its alignment from the vertical to the diagonal while retaining the slight sideways turn. This gives the portrait an aura that might best be characterized as "benignity."

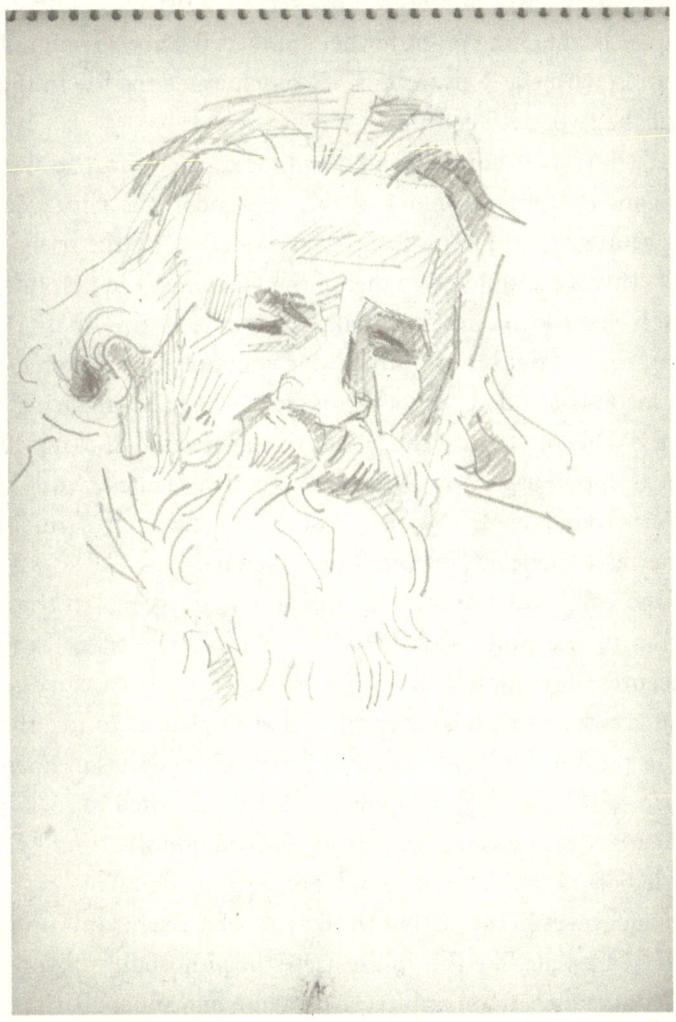

Figure 12c. Preliminary study for the portrait of Gaston Bachelard (23.5 x 15.5 cm). Courtesy Catherine Ballestero.

Unlike the rest of this collection, the portrait (or one of the studies accompanying it) has often been copied, helping to mold the image of Bachelard with his flowing beard, his scarf, and his impressive eyebrows.[12] The beard, surrounded by its ambivalent symbolism of wisdom and genius, served Cristina Chimisso, for example, as a starting point for her study of Bachelard and his confrontation with the conventions of French academic life in the first half of the twentieth century.[13]

If there is a common denominator between the work of Bachelard the philosopher and that of Flocon the engraver, it is what we might term a "philosophy at work" or "on the basis of the work." This is exemplified in one of the *Châteaux*, the first image in a sketch that Flocon headed "make ten castles in the air" (fig. 13a). In fact, that initial series of drawings contains eleven such castles, and Flocon has added another three to the page. Comparing them with the table of plates at the end of the printed volume (fig. 13b), it becomes apparent that not all were actually executed, while other, new ones were added. The castle built on a snail shell is missing, as are the castles labeled "pebbles" and "matches."

The castle on the rock at position 1, relegated to the third position in the final book (fig. 14), received the title "Eyrie" at Bachelard's suggestion.[14] "Rock is a primary image, with an essential role in active and activist literature, that teaches us to live the real in all its profundities and prolixities," Bachelard writes in *Earth and Reveries of Will*, a whole chapter of which is devoted to rock.[15] His commentary on Flocon's fantastical construction site runs: "Truly a page dedicated to the heroism of labor, a page of human geography. Flocon engraves in copper but thinks in rock." Bachelard continues: "Yes, I like a plate like this to illustrate the philosophy of work. It is more concrete than an architect's drawing and more abstract than a finished work. This engraving is a *moment's work*, the moment in which the work advances, the project becomes object, the object takes shape. An abstract-concrete plate."[16]

This expresses one aspect of Flocon's work with great precision. Although Flocon never abandons the figurative scheme, he handles it in a manner so abstract and geometrical that one occasionally seems to lose all orientation. Flocon does not depict; he devises. What is represented is a process. "Yes, indeed,

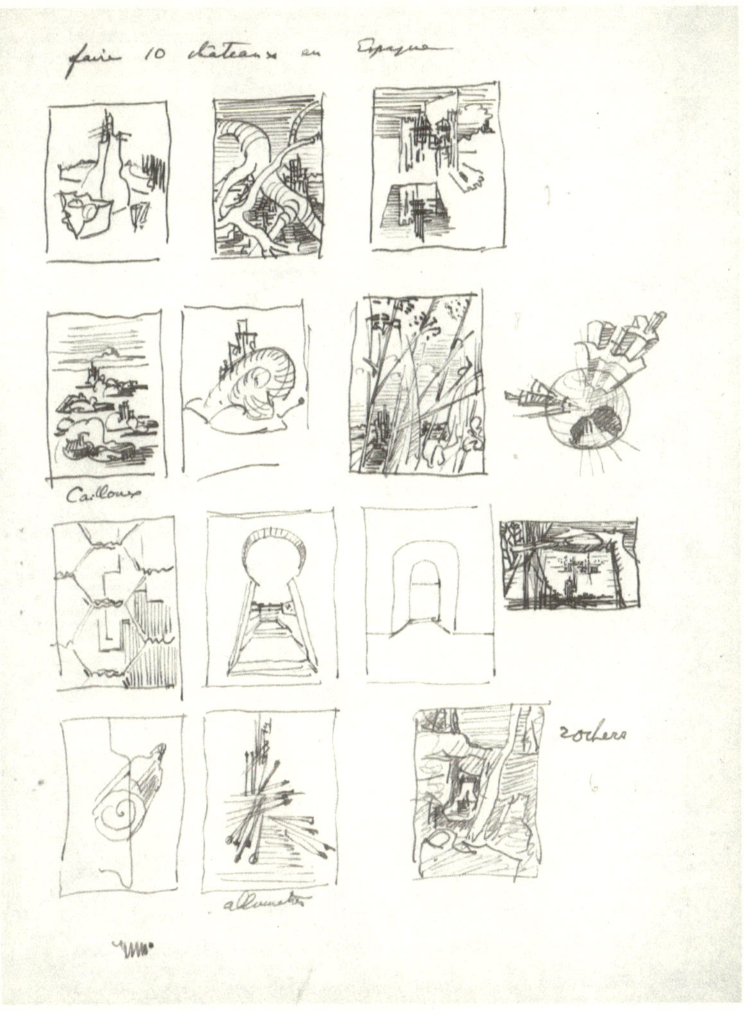

Figure 13a. Sketch entitled "faire 10 châteaux en Espagne" (27.5 x 20.5 cm). Courtesy Catherine Ballestero.

engraving—especially with the burin—is an extremely abstract technique," as Flocon responded to Gil Jouanard's remark that the engraver was left only with abstraction.[17] In turn, Flocon once called Bachelard "the man of the abstract-concrete."[18] He had identified a crucial point in Bachelard's thinking: in his epistemological

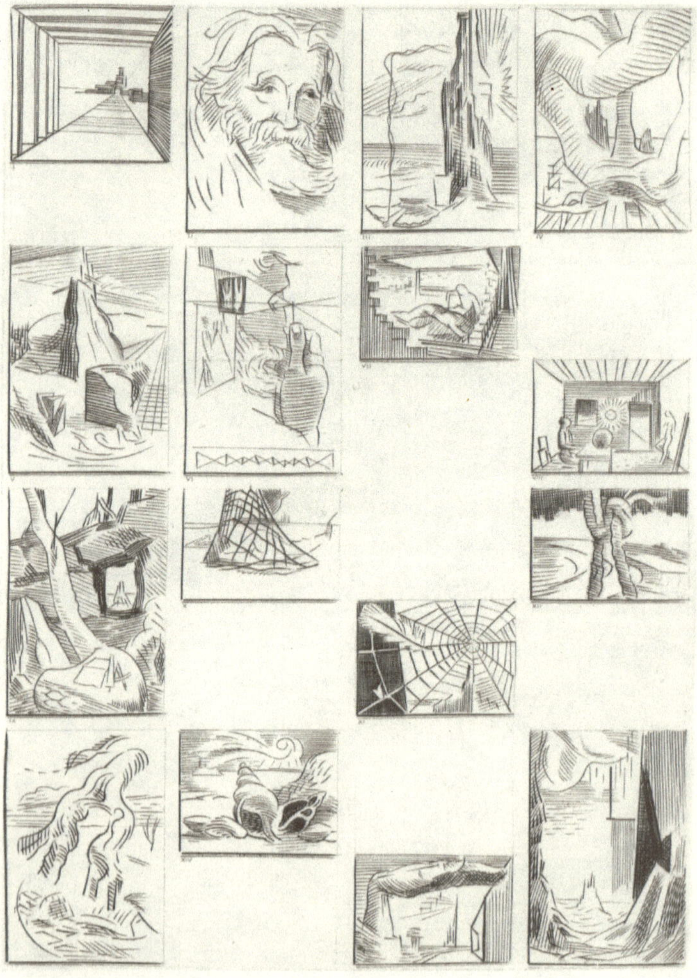

Figure 13b. Table of plates (18 x 13 cm), in Bachelard and Flocon, *Châteaux en Espagne*, 63.

works, Bachelard stresses the conceptual and mathematical nature of thinking in modern physics, yet also relates it to concrete phenomenotechnical arrangements; in his works on poetics, he combines phenomenological concreteness with a free-roaming and imaginative associativity.

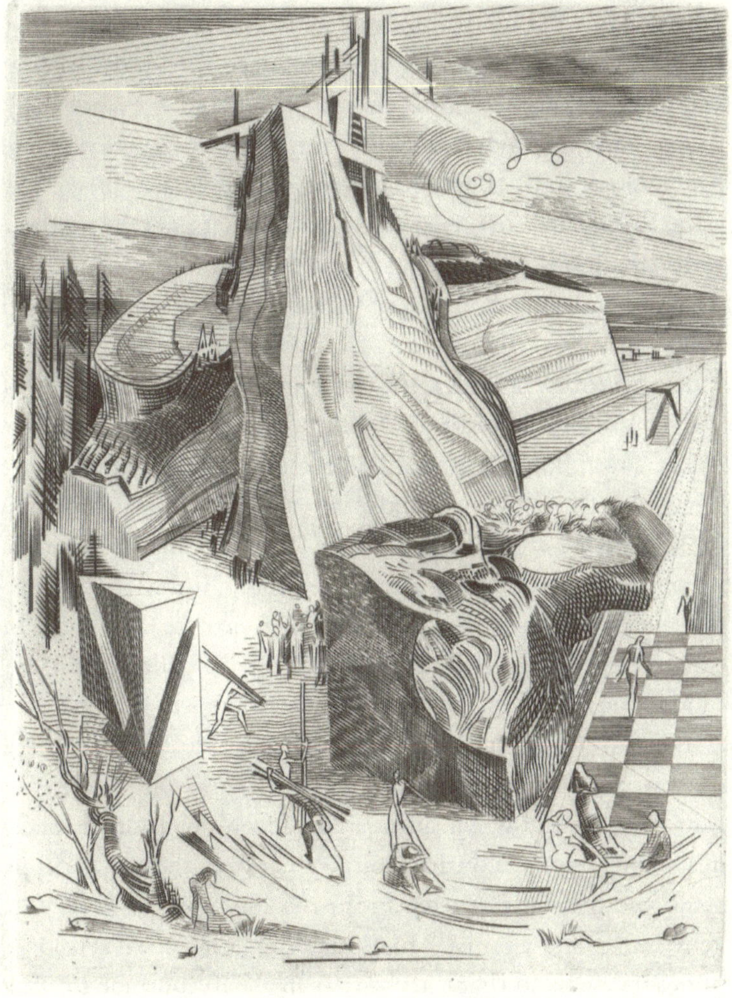

Figure 14. Plate 5, "Le nid d'aigle" (18.5 x 13.5 cm), in Bachelard and Flocon, *Châteaux en Espagne*, 27.

In view of these comments, it will hardly be surprising to find that in *Le rationalisme appliqué*—one of his epistemological works of the second series, around 1950—Bachelard not only attributes an "abstract-concrete mentality" to the physics of his era, but also describes his own philosophy concerned with that mentality as a "philosophy at work" or "philosophy of work" (*philosophie au travail*), as opposed to what he dismissively calls the "philosophies of synopsis" (*philosophies de résumé*) of his predecessors and contemporaries.[19] With that, he claims that his epistemology works in just as process-oriented a manner as do the sciences to which it refers: "Epistemology must therefore be just as mobile as science."[20] For physics, this means that "every new experiment puts the method of experimentation itself to the test";[21] for epistemology, it means that "in order to understand, one must participate in an *emergence*."[22] This is no longer a Cartesian procedure. Events are not determined by simplicity, but by entanglement, complexity, and complicity: "It is no longer a matter of pitting a solitary spirit against an indifferent universe. From now on, it will be necessary to place oneself at the center, where the knowing mind is determined by the precise object of its knowledge and where, in turn, it determines its experience more precisely."[23] For Bachelard, therefore, "in order to keep up with contemporary science, in order to be sensitive to this dynamic of constructed beauty, it is necessary to love difficulty. Difficulty is what gives us awareness of our cultural self."[24]

The commonality with Flocon is palpable here. In the engraver's words, "What drew me to copper engraving was my love of difficulty."[25] And when discussing Flocon's incision, Bachelard writes: "The cut of the burin's point symbolizes the rock drill penetrating the resistant depths. From the moment of its very first, sketchy movement against a hard material, a primeval tension determines in the hand that the entire being shall continue to strain." He adds: "Faced with a hard material, one must continue to act."[26] In both Bachelard's 1948 text on earth and Flocon's 1952 treatise on engraving, labor was present as an *act*; so it is in the castle plates as well, but they go further. In them, labor also acquires vigor and virulence as the *content* of the images.

Here we find science, philosophy, and the arts all lined up together, yet each persists in the singularity of its own particular realizations. For Bachelard, it seems, the work of the engraver instantiates a particular kind of phenomenotechnique. If, as he argues in one of his early essays, modern physics is "no longer a science of *facts*, it is a technology of *effects*" and, no longer content with describing phenomena, is "a production of phenomena,"[27] then the work of the engraver inscribes itself into a comparable space—the space of a "dialectic of nature and anti-nature."[28] This projective element in Flocon's oeuvre is what fascinates Bachelard and resonates so deeply with his own philosophical orientation.

The "castle in the air" shown in figure 14 perfectly epitomizes this dialectic of nature and anti-nature. It reaches up into the clouds. The rock out of which it grows, in vertical and horizontal struts, is hewn like a flint and stands on a flat ground, open to the distance, on which people are working, strolling, conversing, resting. The two heavy masses in front and behind—partly sharp-edged, partly irregular—intensify the flinty impression, yet also cast doubt on it. Are these really rocks? Or are they unevenly split pieces of firewood, laid down or set upright and magnified to a gigantic scale, behind which a thick forest begins? Flocon reveals in his memoirs that he did actually base the image on blocks of oak, firewood for his stove in Bois-le-Roi, near Fontainebleau.[29] He described the evolution of the image in an interview: "I drew very everyday things. For example, I drew some oak logs standing upright, you could see the bark. I used that drawing of logs for my engraving. I made it into very large rocks, and at the very top of the rocks I placed a little construction. For Gaston Bachelard, the logs became a cliff with its eagle's eyrie."[30]

Another striking element of this plate is its strangely inverted relationship of foreground and background. Miniaturized human figures can be seen in the foreground, enormous blocks of half-hewn rock in the background. Into the heart of a classical geometric perspective with its vanishing point on the horizon, Flocon introduces a counterperspective. Thanks to that multiple perspectivity, an engraving like this one can evoke "the greatest contingency of contemplation," comments Bachelard, and he notes, when looking at an image, "I tell myself a different story from the one I told myself yesterday. Sincere contemplation is a capricious thing—pure caprice, in fact."[31] Depending on the perspective taken up, different stories will result, sometimes going their own way and pulling the gaze along with them. This is also how Bachelard sees the scientist's situation in the face of a complex reality. In the sciences, too, "an object may determine several types of objectification, several perspectives of specification, may form part of different problematics."[32]

In the "Eyrie" plate, like so many others in this collection, we also see the workings of yet another game with the gaze, one that goes back to Flocon's training in stage design with Oskar Schlemmer during his Bauhaus years. In a strangely fluctuating triple ambivalence, the engravings combine the space of a landscape, a stage located in that landscape, and a stage decor. It is never quite clear where the landscape ends and the stage begins, or where the stage ends and the decor begins, or indeed the other way around. Moreover, the tiny human figures to be seen on most of the plates tell little additional stories that are only partially connected to the main narrative. They are contingent addenda, specks of reflexivity strewn across the plates that comment on the scene from inside the scene—or, in other cases, seemingly ignore it completely. They bring something accidental to the story. "My books," writes Flocon at the end of the second volume of *Points de fuite*, "almost always revolve around questions raised by the representation (in all the word's senses) of space: the space of life, physico-geometrical space, finally the space of images or of spaces' own mise-en-images; the rich, rich theater of space."[33]

With this in mind, let us turn to the last picture in the *Châteaux* series (fig. 15a). As the closing scene, it is called "The Curtain." Making the theatrical aspect explicit, the curtain falls across the image. The theme of closure is addressed doubly: it is repeated on

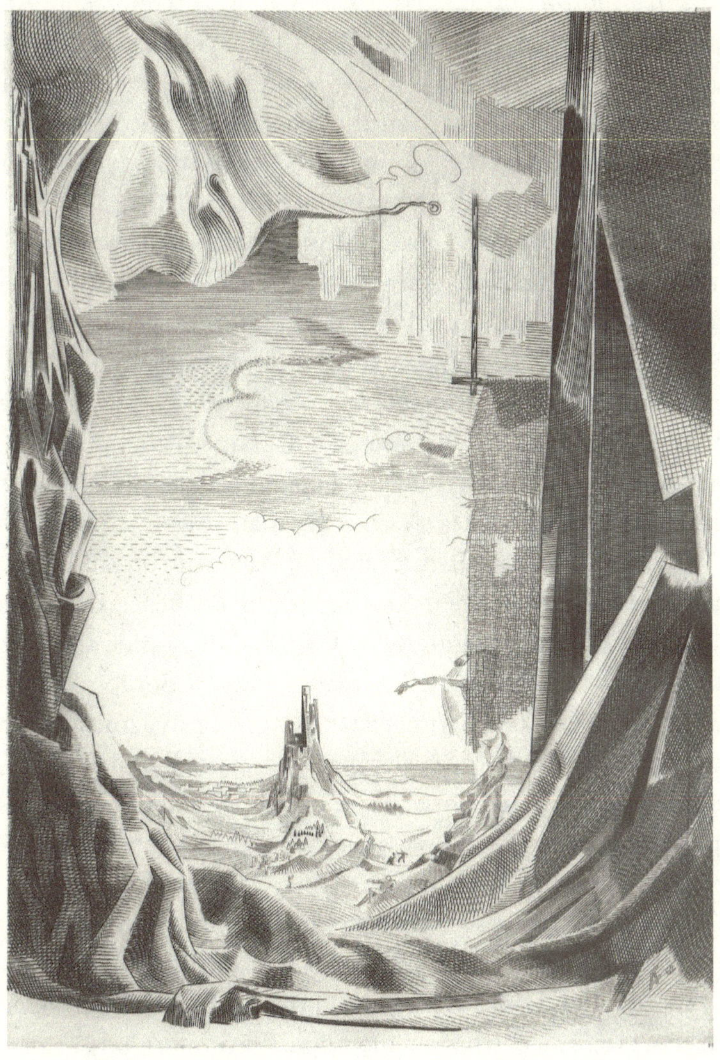

Figure 15a. Plate 16, "Le rideau" (28 x 19.5 cm), in Bachelard and Flocon, *Châteaux en Espagne*, 59.

the right of the image, where a dancer, veiled by coarsely woven gauze—or it might be chainmail—suspended from a rod, makes her parting gesture as she disappears behind the drapery's rock-like folds. In counterpoint, the closing curtain opposite takes on human forms that morph into a tree trunk formation as they ascend. At the picture's upper edge, the curtain takes the shape of fluttering cloths. The half-finished conical tower rearing up at the center of the image also appears to withdraw from the spectator, as do the swarms of birds that are flying off, high up in the sky. Is the dancer waving them good-bye as she stands turned half to the viewer, half to the little figures bustling and gesticulating on the castle grounds? The horizon here is the distant sea—wide, open, and empty, as befits a conclusion that will point beyond its own frontiers. The sense of tension between foreground and distance, surface and depth, is intensified by the image overrunning the margins on three sides of the page. Again we see the interplay of frame, stage, and scenery. They enter into an indissoluble union, and the multiplicity of their mutual determinations leaves space for many divergent narratives.

That multivalence is indicated by Bachelard's reading of the scene, the exact opposite of the one just described: "The castle is built. Now Flocon inaugurates it. He has drawn aside the curtain and the work is revealed in all its theatricality."[34] But for Bachelard, as well, the plate "plays with distance," giving dynamism both to the scene itself and to the eye that contemplates it and evoking the insight "that nothing is fixed for the man who alternates thought with dream." Bachelard closes on a more general note: "Engraving, too, I see now, is an art of mobility."[35]

I would like to look briefly at two of Flocon's drawings for and from this plate. On the preliminary ink sketch (fig. 15b), the frame is only hinted at and the dancer (here in the lower left corner) remains little more than an abstract premonition. The impression of expanse, in contrast, is already fully present in the image. The drawing from the copperplate (fig. 15c) presents a very different impression. Due in large part to the lowered skyline, which in this case blends into the frame, the image suggests that the stage we are looking at is actually the outer framework, not the curtain, which on the right resembles a tower from whose battlements the chainmail curtain hangs, on the left more a garland of plants. The dancer in this drawing, standing

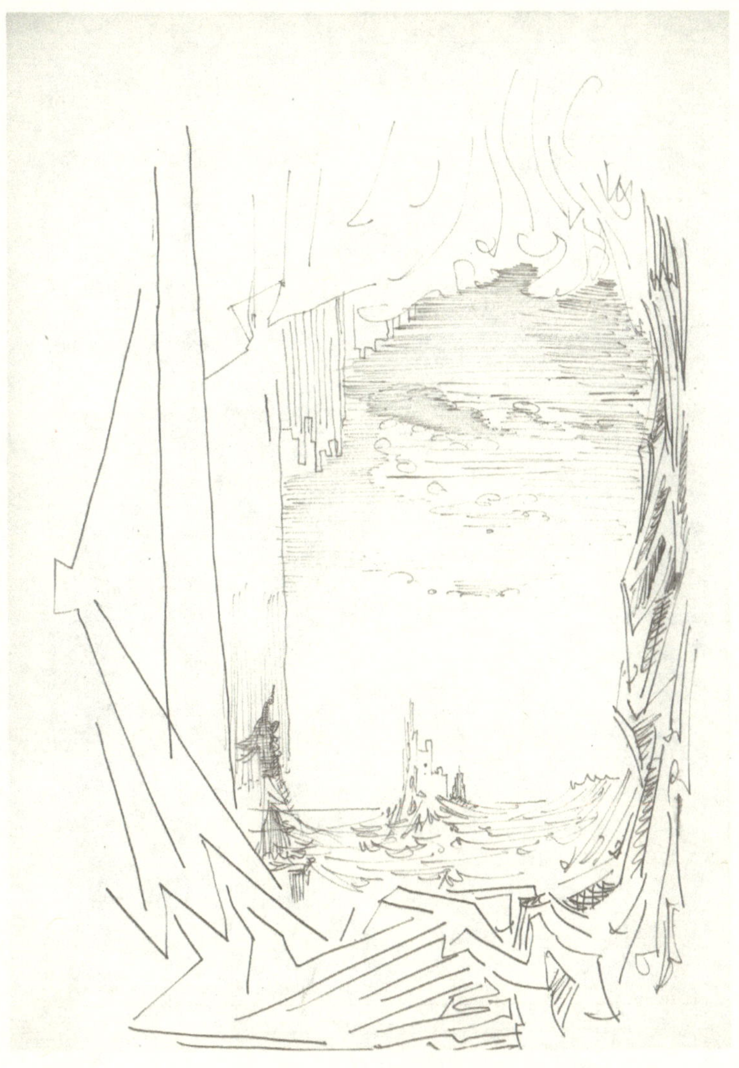

Figure 15b. Ink sketch for "Le rideau" (29 x 20.5 cm). Courtesy Catherine Ballestero.

free on a stepped podium, points the viewer to the castle; the figures in the background have disappeared. Once again, the meaning of the image's components is transformed. Especially the pen-and-ink drawing, which Flocon made as a supplement to one of the books

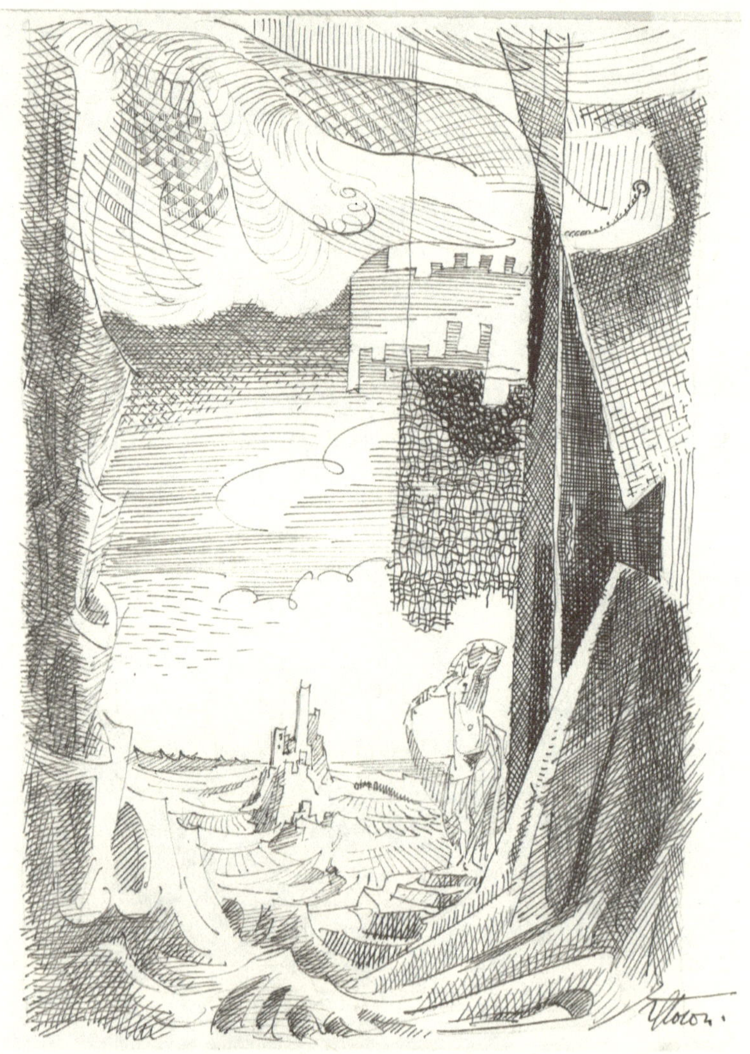

Figure 15c. "Le rideau" (28 x 18 cm), Bachelard and Flocon, *Châteaux en Espagne*, ink drawing inserted into copy no. 72.

and signed for that purpose, supports Bachelard's reading: *Voilà*, the finished castle. On the other hand, Flocon accompanied the drawing with a little picture of a curtain in which the theatrical bow is unambiguously one of farewell (fig. 16). Above all, a comparison with the final engraving indicates the great range of textures and nuances that are possible when the burin gets to work. Demanding the hard edges of metal, they cannot be articulated by the fluidity of ink—however steely the nib—or the softness of graphite, however sharp the pencil.

Flocon worked on the *Châteaux* collection for many years. "The Curtain" dates from 1952, making it one of the first "castles in the air" created for the album. It is therefore noteworthy that the image does not feature at all on the first sketch of the ten castles to be drawn. This underlines the distinction between the order of the plates' making and the order in which they are presented—which was decided retrospectively, as is so often the case for series, whether in the visual arts or in literature. And it reminds us that in the sciences, too, the registers of research differ from the orders of representation; the two cannot be aligned.

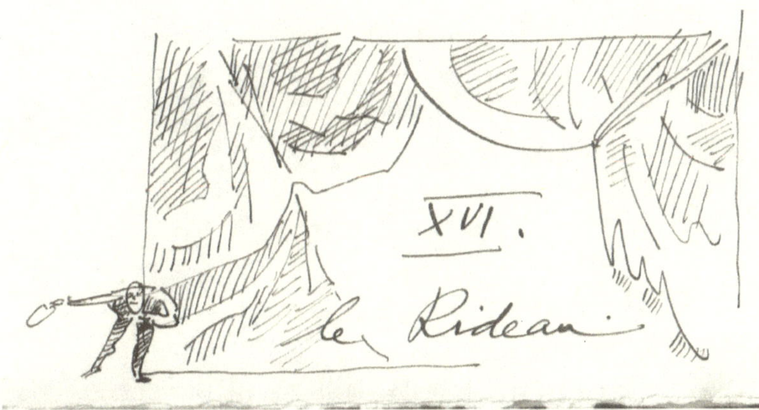

Figure 16. Bachelard and Flocon, *Châteaux en Espagne*, presentation of the drawings (6 x 11 cm), inserted into copy no. 72.

Enclosed with the copy of the book that I examined was the solid copper plate on which this engraving was executed. The plate bears the stamp of the copper polisher Cossonneau at 33, rue Descartes. Flocon used secondhand nineteenth-century printing plates for the *Châteaux* works. He had them scraped and polished, and he took account of their various sizes when he formatted the designs.[36] Looking at the engraved plate, unlike looking at the prints, means taking it in one's hands, turning it and playing with the light that falls on the metal—it is impossible to get a view of the incisions across the whole plate at once. Not least, a printing plate of this kind reminds us that as the engraver goes about his work, holding the burin in one hand and turning the plate in the other, he is moving in a different space. The fact that the image is reversed is the most trivial aspect here. An unpracticed observer would never be able to see in the emerging scratches on the metal what will later become visible in the print. To live and transform this hiatus, to drive tracks into the plate with the precision that the copper commands, in other words to obey its hard and implacable metallic materiality while simultaneously anticipating the pressure onto soft, lush, absorbent paper: that is the irreducible tension peculiar to the engraver's craft.

SCIENCE, ART, LITERATURE

During the years of his contact with Flocon, Bachelard on many occasions took up the pen as a "philosopher of images" and described himself as such.[1] As early as 1945, he commented on José Corti's *Rêves d'encre* with a short text entitled "Une rêverie de la matière," in which, again, Bachelard examines the authority of a material, this time ink: "Corti really has bowed to the black liquid's will; and in the deepest part of that will he has experienced some mysterious nostalgia of iron and alum both yearning to expand, struggle, strike up a partnership, live again, proliferate, jostle one another, create."[2] In 1946, he completed a text for *Les devins*, a collection of sixteen drypoint etchings by the graphic artist and painter Louis Marcoussis, who had died five years earlier.[3] In the year of his own death, 1962, Bachelard wrote another short text for an homage to Marcoussis presented by Galerie Kriegel, Paris. In this, he describes himself as a "friend of painters and poets."[4] The essay's title, "Marcoussis et ses poètes," recalls that Marcoussis, strongly influenced by cubism, had worked on book projects in the 1920s and 1930s with writers including Tristan Tzara.[5]

Other short publications testify to Bachelard's continuing contact with artists from Flocon's circle, especially Christine Boumeester. For the first volume of the series "Art de demain," on Boumeester, Bachelard wrote the text "Gravure et profondeur" at Flocon's request.[6] "My dear friend," a letter from Bachelard to Flocon dated October 23, 1951, begins, "I have written two pages for Christine Boumeester. Before I copy them, I would like to show them to you and talk to you about them."[7] The piece was reprinted ten years later in an exhibition catalog.[8] In 1953, Bachelard wrote a text for Jacques Fouquet,[9] and in 1956, for Maeght's well-known periodical *Derrière le miroir*, he commented on works in iron by the Basque sculptor Eduardo Chillida, who had lived in Paris from 1948 to 1951.[10] In 1960, he supplied a text for a short monograph on the tapestries of Asger Jorn and Pierre Wemaëre.[11] The same year, he wrote the introduction to one of the two volumes of Marc Chagall's *Dessins pour la Bible* published in the Verve series, the one containing Chagall's drawings and lithographs of 1958 and 1959.[12]

In turn, the artist Flocon sought direct dialog with the natural sciences of his day on many occasions. A bibliophile edition of Jean Rostand's *Notes d'un biologiste* appeared in 1954,[13] where the text preceded the plates, unlike in the books Flocon made with Bachelard. Flocon had met the biologist Rostand in 1950, through Bachelard, and visited his private laboratory and vivarium in Ville-d'Avray with its monstrous frogs. "How beautiful the interiors of the vivarium are! And what labyrinths! You will bring this mystery fully to light," Bachelard wrote to Flocon.[14] In the resulting "engraved reportage,"[15] Flocon drew inspiration primarily from the "simple material of the laboratory" and from the "animal-monsters," making Rostand's frogs the principal protagonists.[16] In 1956, already teaching at the art and design school École Estienne in Paris, Flocon created wood engravings for an unpublished text by the physicist Louis de Broglie.[17] Finally, he also carried out historical work. He published a history of the book in 1961[18] and around the same time collaborated with the mathematician and historian of mathematics René Taton (incidentally another contact made through Bachelard[19]) on a study of the history and practices of perspective.[20] This set the stage for Flocon's later theoretical works on curvilinear perspective. Around the same time, he began an extensive correspondence with Maurits Cornelis Escher, who regarded Flocon as a "brother in spirit," as he put it in one letter.[21] Later, Flocon dedicated a whole treatise, written with his colleague André Barre of the École Estienne, to the representation of curvilinearity.[22]

Hedwig Wingler once called Flocon an "artist-researcher-historian,"[23] and Yves Michaud described him as someone who "locates art on the side of knowledge rather than on the side of inspired genius": "He doesn't do just anything—he thwarts constraints by going right up to the boundaries of his profession's possibilities."[24] Flocon saw himself the same way: "I loved to push rationality to its very limits: beyond that, quite enough mystery remains."[25] It would be difficult to express more cogently the stance of an experimenter. The experimenter organizes knowledge in such a way that he or she becomes able to exceed it. Experimentation is the craft of the abstract-concrete.

Accordingly, Bachelard perceived Flocon's works as variations on an enduring effort to abstract from the concrete while concretizing the abstract. Just as he was interested in the Real of literature, not the stories that it told, so he was interested in the Real of engraving and its iterative potential. *Earth and Reveries of Will* had sparked the shared interest between Bachelard and Flocon. In it, Bachelard reviewed literary images of the earth and of the resistance that earth offers in both hard and pliable forms, starting from the precept that literature must surprise. Bringing "new life to language by creating new images," he writes, is "the function of literature and poetry."[26] There follows a judgment on one of the most characteristic literary currents of his era: "Certain arenas of contemporary literature seem marked by a virtual *explosion of language*. . . . In a more liberated poetry, surrealist poetry for example, language comes to full efflorescence."[27] It is against the material of language that the labor of poetry and the labor of reading chafe, drive, outdo themselves: "I am a dreamer of words, of written words. I think I am reading; a word stops me. I leave the page. The syllables of the word begin to move around. Stressed accents begin to invert."[28]

This autonomy is also what attracts Bachelard to the natural sciences and to the work of artistic engraving. His fascination with the dynamics of objectification—and the phenomenotechniques of knowing, saying, and showing that it deploys—gives his work its inner unity. In his own oeuvre, Bachelard articulates that unity with an unfurling into epistemology, imagology, and finally poetics. For even if they all share one thing, namely, that they emerge through the resistance of their materials, each such material follows its own imponderables, its own possibilities of engraftment. Literary expression, Bachelard writes in *Earth and Reveries of Will*, has an "independent existence,"[29] and this existence is what he sought in his voracious reading. Bachelard's studies of graphic expression, in contrast, were most lastingly marked by his encounter with Flocon.

In *Water and Dreams*, Bachelard used the notion of the graft to express this material agency of language. The graft adds something new to the stock, something not derivable from it. Yet for its own leaves to bud, the graft is dependent on the stock. In Bachelard's view, the graft is "not simply a metaphor," but a figure of material imagination: "It is the graft that can truly provide the material imagination with an exuberance of forms, that can transmit the richness and density of matter to formal imagination." He adds, almost categorically: "By nature, art is grafted."[30]

But in all three fields—science, literature, art—within which their works move, Bachelard and Flocon are interested less in the "life of forms" (to recall once more the title of Focillon's book), in the sense of pure form, than in the life of particular instances of *becoming*-form, instances provoked by matter. The philosopher was fascinated by the engraver's craft, which, in its encounter with matter, seemed to combine phenomenotechnical with poetic elements. The engraver was fascinated by a poetologist who mined his images not from introspection but from the elements, and who celebrated the scientific approach to the world as a breach with everyday experience. Both were fanatically concerned with detail. If Bachelard once called himself a "philosopher of micro-epistemology,"[31] Flocon might be described as a bricoleur of the grapheme, an artist of the elemental gesture of line. One in the realm of writing, the other in the realm of graphic art, both men confronted the epistemological, poetic, and artistic possibilities of precision—of sharpening tools, sharpening things, a process that as we know may bring either innovation or breakage. For them, precision was a permanent challenge, a movement that ultimately finds its fulfillment only in iteration.

NOTES

INTRODUCTION

1. Albert Flocon, "Verflechtungen," in *Prägungen: Deutsche in Paris*, ed. Georg Lechner, 284–98 (Düsseldorf: Bollmann, 1991), here 285. [Translator's note: Here and throughout, all translations are my own unless otherwise attributed.]

2. In the German-speaking world, Albert Flocon is almost unknown outside those fields as well. One exception is a bilingual presentation of his experimental prints compiled by the Bauhaus Archive: Albert Flocon, *Suites expérimentales, 1943–1983* (Vienna: Medusa, 1983). Another bilingual publication is the catalog accompanying the 1992 exhibition at the École des Beaux-Arts de Metz: *Albert Flocon: Une poétique de la vision. Du Bauhaus à la perspective curviligne / Albert Flocon: Eine Poetik des Sehens. Vom Bauhaus zur kurvenlinearen Perspektive* (Metz: École des Beaux-Arts de Metz, 1992). Stefan Bollmann has published a compilation from his books, with a brief introduction: Paul Éluard, Gaston Bachelard, and Albert Flocon, *Die Bücher des Albert Flocon*, trans. Nicolaus Bornhorn (Düsseldorf: Bollmann, 1991). In his important biography of Gaston Bachelard, the French literary and art critic André Parinaud has just two peripheral pages to spare for the philosopher's collaboration with Flocon. André Parinaud, *Gaston Bachelard* (Paris: Flammarion, 1996), 414–15.

SURRATIONALISM

1. Gaston Bachelard, "Surrationalism," trans. Julien Levy, in *Surrealism*, ed. Julien Levy (New York: Black Sun Press, 1936), 186–89,

here 186 (translation emended; here and in all quotations, italics are original). The French text appeared as "Le surrationalisme," in *Inquisitions: Organe du groupe d'études pour la phénoménologie humaine* 1 (June 1936): 1–6. This was the first and only issue of the journal, which arose in the milieu of the Popular Front.

2. Tristan Tzara, *Grains et issues* (1935), edited with an introduction and notes by Henri Béhar (Paris: Garnier-Flammarion, 1981).

3. Tzara, *Grains et issues*, 155.

4. Tzara, *Grains et issues*, 155–56.

5. Tzara, *Grains et issues*, 157.

6. Tzara, *Grains et issues*, 157.

7. Gaston Bachelard to Roger Caillois, November 2, 1935. Reproduced in Henri Béhar (ed.), *Du Surréalisme au Front Populaire: Inquisitions* (Paris: Éditions du CNRS, 1990), 152.

8. Jean Wahl, Review of *Inquisitions*, *La nouvelle revue française* 275 (August 1, 1936): 402–03, reproduced in Béhar, *Surréalisme*, 169.

9. Tzara, *Grains et issues*, 156.

10. Gaston Bachelard, *The New Scientific Spirit*, trans. Arthur Goldhammer (Boston: Beacon Press, 1984), 135, 173 (*Le nouvel esprit scientifique*, 1934).

11. Bachelard, *New Scientific Spirit*, 176.

12. Bachelard, *New Scientific Spirit*, 168.

13. Bachelard's text appeared as "Germe et raison dans la poésie de Paul Éluard," *Europe* 31 (1953): 115–19. It is translated as "Germ and Reason in the Poetry of Paul Eluard," in Gaston Bachelard, *The Right to Dream*, trans. J. A. Underwood (New York: Orion, 1971), 145–52, here 145.

14. Bachelard, "Surrationalism," 188 (translation emended).

ALBERT FLOCON, GASTON BACHELARD

1. Albert Flocon, *Points de fuite, 1: 1909–1933* (Neuchâtel: Éditions Ides et Calendes, 1994), 166.

2. Albert Flocon, "Main ouvrière et main rêveuse (Entretien avec Gil Jouanard)," in "Bachelard ou le droit de rêver," special issue, *Solaire* 10 (1983): 57–65, here 65.

3. Wassily Kandinsky, *Point and Line to Plane* (1926), trans. Howard Dearstyne and Hilla Rebay (New York: Solomon R. Guggenheim Foundation, 1947).

4. See, for example, Karin von Maur, "Oskar Schlemmer as Choreographer and Stage Designer," in *Oskar Schlemmer: Visions of a New World*, ed. Ina Conzen and Staatsgalerie Stuttgart (Munich: Hirmer, 2014), 191–246, plate 13.

5. See Wulf Herzogenrath, "Von der Bauhaus-Bühne zum kurvenlinearen Raumbild: Albert Flocon," in *Das Bauhaus und Frankreich: Le Bauhaus et la France. 1919–1940*, ed. Isabelle Ewig, Thomas W. Gaehtgens, and Matthias Noell (Berlin: Akademie, 2002), 465–78.

6. Albert Flocon, *Scénographies au Bauhaus, Dessau, 1927–1930: Hommage à Oskar Schlemmer en plusieurs tableaux* (Paris: Archimbaud, 1987).

7. Flocon, *Points de fuite, 1*, 252.

8. Albert Mentzel and Albert Roux, eds., *Formes nues* (Paris: Éditions d'Art Graphique et Photographique, 1935); in the "Père Castor" series for children: Mentzel, *Villages de France: Jeu de construction* (Paris: Flammarion, 1937); Mentzel, *Le cirque animé* (Paris: Flammarion, 1938).

9. Albert Flocon, *Points de fuite, 2: 1933–1994* (Neuchâtel: Éditions Ides et Calendes, 1995), 109.

10. Ferdinand Flocon (1800–1866) was a radical liberal journalist and writer and a member of the Provisional Government of the Second French Republic of 1848. After the coup d'état of 1851, he went into exile in Lausanne, Switzerland.

11. See Parinaud, *Gaston Bachelard*, chapter 1.

12. Gaston Bachelard, *Water and Dreams: An Essay on the Imagination of Matter*, trans. Edith R. Farrell (Dallas: The Dallas Institute of Humanities and Culture, 1983), 7–8 (*L'eau et les rêves: Essai sur l'imagination de la matière*, 1942).

13. Bachelard to "Monsieur" [Gustave Le Bon], January 18, 1931, from Dijon (private collection).

14. Gaston Bachelard, *Essai sur la connaissance approchée* (Paris: Vrin, 1928); Bachelard, *Étude sur l'évolution d'un problème de physique: La propagation thermique dans les solides* (Paris: Vrin, 1928).

15. Bachelard, *New Scientific Spirit*; Gaston Bachelard, *The Formation of the Scientific Mind: A Contribution to a Psychoanalysis of Objective Knowledge*, trans. Mary McAllester Jones (Manchester: Clinamen, 2002) (*La formation de l'esprit scientifique*, 1938).

16. Gaston Bachelard, *The Psychoanalysis of Fire*, trans. Alan C. M. Ross (Boston: Beacon Press, 1964) (*La psychanalyse du feu*, 1938).

17. Bachelard to Marius Filloux, May 10, 1939, reproduced in Jean-Claude Filloux, "Témoignage sur la vie de Gaston Bachelard," in *Gaston Bachelard: Science et poétique, une nouvelle éthique?*, ed. Jean-Jacques Wunenburger (Paris: Hermann, 2013), 493–502, here 498.

18. Gaston Bachelard, *The Philosophy of No: A Philosophy of the New Scientific Mind*, trans. G. C. Waterston (New York: Orion Press, 1968) (*La philosophie du non*, 1940).

19. Bachelard, *Water and Dreams*; Gaston Bachelard, *Air and Dreams: An Essay on the Imagination of Movement*, trans. Edith R. Farrell and C. Frederick Farrell (Dallas: The Dallas Institute of Humanities and Culture, 1988) (*L'air et les songes: Essai sur l'imagination du mouvement*, 1943); Bachelard, *Earth and Reveries of Will: An Essay on the Imagination of Matter*, trans. Kenneth Haltman (Dallas: The Dallas Institute of Humanities and Culture, 2002) (*La terre et les rêveries de la volonté: Essai sur l'imagination des forces*, 1948); Bachelard, *Earth and Reveries of Repose: An Essay on Images of Interiority*, trans. Mary McAllester Jones (Dallas: The Dallas Institute of Humanities and Culture, 2011) (*La terre et les rêveries du repos: Essai sur les images de l'intimité*, 1948).

PERSPECTIVES

1. Flocon, *Points de fuite*, 2, 131.
2. Flocon, *Points de fuite*, *2*, 134.

3. Paul Éluard and Albert Flocon, *Perspectives: Poèmes sur des gravures de Albert Flocon* (Paris: Maeght, 1948). See also Jean-Charles Gateau, *Éluard, Picasso et la peinture, 1936–1952* (Geneva: Librarie Droz, 1983), 211–17.

4. Flocon, *Points de fuite, 2*, 136–38.

5. Éluard and Flocon, *Perspectives*, plate 1.

6. Henri Focillon, "In Praise of Hands" (1939), trans. S. L. Faison, in Focillon, *The Life of Forms in Art*, trans. Charles Beecher Hogan, George Kubler, and S. L. Faison (New York: Zone Books, 1989), 157–85.

7. See the photograph in Jean-Claude Margolin, *Bachelard* (Paris: Seuil, 1974), 23.

8. Focillon, "In Praise of Hands," 158.

9. Focillon, "In Praise of Hands," 166.

10. Volume 2 of *La table ronde*, which appeared in Paris in April 1945, includes Henri Focillon's "Éloge des lampes" and a facsimile of Georges Auric's setting of a manuscript poem by Paul Éluard.

11. Martin Warnke has linked the hand's prominence in the first half of the twentieth century to the rediscovery of Mannerism. Warnke, "Der Kopf in der Hand," in Martin Warnke, *Nah und Fern zum Bilde. Beiträge zu Kunst und Kunsttheorie*, ed. Michael Diers (Cologne: Dumont, 1997), 108–20, here 119.

12. Man Ray and Paul Éluard, *Les mains libres: Dessins de Man Ray illustrés par les poèmes de Paul Éluard* (Paris: Éditions Jeanne Bucher, 1937).

13. "Les mains de Paul Éluard," photograph by Dorka Raynor, 1948, reproduced in a Librairie Jean-Yves Lacroix catalog, 2013.

14. Denis de Rougemont, *Penser avec les mains* (Paris: Albin Michel, 1936), 147, 152.

15. Focillon, "In Praise of Hands," 166. For the case of nineteenth-century astronomy, see, for example, Omar W. Nasim, *Observing by Hand: Sketching the Nebulae in the Nineteenth Century* (Chicago: University of Chicago Press, 2013).

16. Anne Vernay and Richard Walter, eds., *La main à plume: Anthologie du surréalisme sous l'occupation*, preface by Gérard Durozoi

(Paris: Syllepse, 2008). The title phrase is from "Mauvais sang" in Rimbaud's *Une saison en enfer*: "La main à plume vaut la main à charrue" (the hand at the pen is as valuable as the hand at the plow).

17. Flocon, *Points de fuite, 1*, 167.
18. Éluard and Flocon, *Perspectives*, plate 10.
19. Flocon, *Points de fuite, 1*, 346.

REVERIES OF EARTH

1. Bachelard, *Earth and Reveries of Will*; Bachelard, *Earth and Reveries of Repose*.
2. Bachelard, *Water and Dreams*; Bachelard, *Air and Dreams*.
3. Gaston Bachelard, *The Flame of a Candle*, trans. Joni Caldwell (Dallas: The Dallas Institute of Humanities and Culture, 1988) (*La flamme d'une chandelle*, 1961).
4. Gaston Bachelard, *Fragments of a Poetics of Fire*, trans. Kenneth Haltman (Dallas: The Dallas Institute of Humanities and Culture, 1990) (*Fragments d'une poétique du feu*, 1988).
5. Bachelard, *Water and Dreams*, 3.
6. Bachelard, *Earth and Reveries of Will*, 7.
7. Bachelard, *Earth and Reveries of Will*, 16–17.
8. Bachelard, *Earth and Reveries of Will*, 19.
9. Bachelard, *Earth and Reveries of Will*, 24.
10. Bachelard, *Earth and Reveries of Will*, 37.
11. Paul Éluard, *Le livre ouvert II* (Paris: Éditions Cahiers d'Art, 1942), 121; English translation in Bachelard, *Earth and Reveries of Will*, 37.
12. Bachelard, *Earth and Reveries of Repose*, 2.
13. Bachelard, *Earth and Reveries of Repose*, 2.

IN PRAISE OF HANDS

1. Flocon, *Points de fuite, 2*, 139.

2. Flocon, *Points de fuite, 2*, 143.

3. Flocon, *Points de fuite, 2*, 144. For more detail on this meeting, see Flocon, "Main ouvrière et main rêveuse," 57–58.

4. Gaston Bachelard, Paul Éluard, Jean Lescure, Henri Mondor, Francis Ponge, René de Solier, Tristan Tzara, and Paul Valéry; engravings by Christine Boumeester, Roger Chastel, Pierre Courtin, Sylvain Durand, Jean Fautrier, Marcel Fiorini, Albert Flocon, Henri Goetz, Léon Prébandier, Germaine Richier, Jean Signovert, Raoul Ubac, Roger Vieillard, Jacques Villon, Gérard Vulliamy, and Albert-Edgar Yersin, *À la gloire de la main* (Paris: Aux dépens d'un amateur, 1949). Bachelard's contribution, "Matière et main," is translated as "Hand vs. Matter" in Bachelard, *Right to Dream*, 56–58.

5. Flocon, *Points de fuite, 2*, 141–42.

6. Flocon, *Points de fuite, 2*, 142.

7. Flocon, *Suites expérimentales*, 17.

8. Bachelard et al., *À la gloire de la main*, plate 7.

9. Tristan Tzara, "Poème," in Bachelard et al., *À la gloire de la main*, 45.

10. Albert Flocon, "Le philosophe et le graveur," in *Bachelard, Colloque de Cerisy* (Paris: Union Générale d'Éditions, 1974), 271–78, here 272.

11. Bachelard, "Hand vs. Matter," 56–57.

12. Paul Valéry, ". . . manuopera, manœuvre, œuvre de main," in Bachelard et al., *À la gloire de la main*, 51.

13. Stefan Bollmann, "Vorwort," in Éluard, Bachelard, and Flocon, *Die Bücher des Albert Flocon*, 7–8.

14. Bachelard, "Hand vs. Matter," 57.

15. Bachelard, "Hand vs. Matter," 58.

16. Flocon, *Points de fuite, 2*, 144.

17. Flocon, "Le philosophe et le graveur," 275.

18. Flocon, *Points de fuite, 2*, 190.

19. Flocon, "Main ouvrière et main rêveuse," 59.

20. Flocon, *Suites expérimentales*.

21. Albert Flocon, "Regard en arrière," in *Albert Flocon: Une poétique de la vision*, 77–83, here 83.

22. Albert Flocon, "Qu'est-ce qu'une suite expérimentale?," in Flocon, *Suites expérimentales*, 19.

23. Georges Didi-Huberman, *La ressemblance par contact: Archéologie, anachronisme et modernité de l'empreinte* (Paris: Minuit, 2008), 33.

LANDSCAPES

1. Gaston Bachelard and Albert Flocon, *Paysages* (Rolle: Eynard, 1950), reprinted in 1982 by Éditions de l'aire, Lausanne. Bachelard's introductory text, "Introduction à la dynamique du paysage," is translated as "Introduction to the Dynamics of Landscape" in Bachelard, *Right to Dream*, 59–78.

2. "Des extraits par les auteurs, d'un livre à paraître aux éditions Eynard (Rolle, Suisse): Paysages, notes d'un philosophe pour un graveur," *Cobra* 6 (April 1950): 15.

3. Flocon, "Le philosophe et le graveur," 273.

4. Flocon, *Points de fuite, 2*, 181.

5. Flocon, *Points de fuite, 2*, 181.

6. A frontispiece with this title precedes the "suite des 16 burins, avec remarques, sur Chine," which was enclosed in the twenty-four copies of the deluxe edition.

7. Bachelard to Flocon, Dijon, July 20 (or July 30; numbers overwritten), 1949. This and the following letters were inserted into copy no. 7 of *Paysages*.

8. Bachelard to Flocon, Dijon, August 4, 1949.

9. Flocon, "Le philosophe et le graveur," 275.

10. Bachelard to Flocon, Paris (clearly a slip, as the address given in the letter is Dijon), August 8, 1949.

11. Bachelard, "Introduction to the Dynamics of Landscape," 59.

12. Bachelard, "Introduction to the Dynamics of Landscape," 59.

13. Bachelard, "Introduction to the Dynamics of Landscape," 64.

14. Pascal Fulacher, "Albert Flocon: Graveur et bibliomane," *Art et métiers du livre* 135 (June 1985): 33–36, here 34.

15. Bachelard and Flocon, *Paysages*, 7.

16. Flocon, *Points de fuite, 2*, 182.

17. Flocon, "Le philosophe et le graveur," 276.

18. Bachelard, "Introduction to the Dynamics of Landscape," 71.

19. Bachelard, "Introduction to the Dynamics of Landscape," 60. On the theme of force, provocation, and will, see also Valeria Chiore, "Force, provocation, volonté: *Paysages. Notes d'un philosophe pour un graveur*, entre ontologie des éléments et phénoménologie de la parole poétique," *Ideação* 25, no. 1 (2011): 121–36; and Valeria Chiore, "Bachelard-Flocon, *Paysages*: Il mouvement—Forza, Provocazione, Volontà tra *éléments* e *poésie*," in "Bachelard e la plasticità della materia," special issue, *Altre Modernità* (2012): 76–88.

20. Jean Starobinski, "Paysages gravés," *Journal de Genève* (January 27, 1951): 3–4, here 4.

21. Flocon, *Points de fuite, 2*, 181.

22. Flocon, *Points de fuite, 2*, 181.

23. "A Flocon 51." Flocon returned to this motif several times, for example, in a late wood engraving not reproduced here, but to be found in the supplement to *Graphische Kunst (Zeitschrift für Graphikfreunde)* 32, no. 1/A (1989) (Memmingen: Edition Curt Visel, 1989).

24. Flocon, "Main ouvrière et main rêveuse," 61.

25. Bachelard, "Introduction to the Dynamics of Landscape," 60.

26. Flocon, *Points de fuite, 2*, 181–82.

27. Flocon, *Points de fuite, 2*, 181.

28. "Le philosophe et le graveur: Entretien entre Gaston Bachelard, Albert Flocon et Jean Amrouche (mai 1950, mai 1953)," excerpted in *Bulletin de l'Association des Amis d'Albert Flocon* (1999): 4–18, here 9.

29. These five letters from Bachelard to Flocon were sent from Dijon in summer and early fall 1949. Catherine Ballestero kindly gave me access to five further letters from Bachelard, sent to Flocon from Dijon between August 1949 and September 1958, from Flocon's papers (Catherine Ballestero's archives).

30. Bachelard to Flocon, Paris [i.e., Dijon], August 8, 1949.

31. Bachelard to Flocon, Dijon, August 18, 1949. Bachelard may have been thinking of his interest in the graphic work of Louis Marcoussis. Louis Marcoussis, *Les devins: 16 Pointes sèches*, text by Gaston Bachelard, preface by Maurice Raynal (Paris: La Hune, 1946).

32. Bachelard to Flocon, Dijon, August 18, 1949. Chronologically, this must have been *L'activité rationaliste de la physique contemporaine*, which appeared in the first third of 1951 with Presses Universitaires de France in Paris. It was preceded in 1949 by *Le rationalisme appliqué* and followed in 1953 by *Le matérialisme rationnel*, both published by Presses Universitaires de France, Paris. In other words, at the time Bachelard was working in parallel on the second part of his late epistemological trilogy and on a poetics of the visual.

33. Bachelard to Flocon, Dijon, August 26, 1949.

34. Bachelard to Flocon, Dijon, October 2, 1949.

35. They are enclosed in copy no. 12 of *Paysages*, the one held by the Library of the Max Planck Institute for the History of Science.

36. Bachelard, "Introduction to the Dynamics of Landscape," 68.

37. Bachelard, "Introduction to the Dynamics of Landscape," 68.

38. Bachelard, "Introduction to the Dynamics of Landscape," 70.

39. Bachelard, "Introduction to the Dynamics of Landscape," 69.

40. "Le philosophe et le graveur: Entretien entre Gaston Bachelard, Albert Flocon et Jean Amrouche," 6.

41. Gottfried Boehm, "Bildsinn und Sinnesorgane," *Neue Hefte für Philosophie* 18/19 (1980): 118–32, here 128; see also Robert Zwijnenberg, "Rafaël zonder handen: Enige voorbereidende opmerkingen over een esthetica met handen," *Kunstlicht* 17, no. 1 (1996): 23–28.

42. Rodolphe Gasché contributing to the "Roundtable on Translation," in *The Ear of the Other: Otobiography, Transference, and Translation. Texts and Discussions with Jacques Derrida*, ed. Christie McDonald, trans. Peggy Kamuf and Avital Ronell (Lincoln: University of Nebraska Press, 1988), 93–161, here 113–14.

43. Bachelard, *Earth and Reveries of Repose*, 214.

44. Bachelard, *Earth and Reveries of Repose*, 214.

45. Bachelard to Flocon, Dijon, September 8, 1950. Bachelard writes "péché" (sin) instead of "pêcher" (peach tree). Flocon added an asterisk to the word and noted "nice slip."

ON ENGRAVING

1. Albert Flocon, *Traité du burin. Illustré par l'auteur. Préface de Gaston Bachelard* (Paris: Librairie Auguste Blaizot, 1952). The essay was published as a booklet in 1954 by Éditions Pierre Cailler in Geneva and reprinted in 1981, with retrospective commentary by the author, by Clancier/Guénaud of Paris. Bachelard's preface, "Le *Traité du Burin* d'Albert Flocon," is translated as "Albert Flocon's 'Engraver's Treatise,'" in Bachelard, *Right to Dream*, 79–82.

2. Flocon, *Points de fuite, 2*, 184.

3. Flocon, "Regard en arrière," 82.

4. Albert Flocon, "L'éloge du burin," *Art d'aujourd'hui* 9 (April 1950): 13.

5. Such is the title of chapter 1 in Bachelard's *Formation of the Scientific Mind*.

6. Flocon, *Traité du burin*, 24.

7. Flocon, "Éloge du burin," 13.

8. Flocon, "Regard en arrière," 81.

9. Flocon, *Traité du burin*, 24.

10. Flocon, *Suites expérimentales*, 11.

11. Flocon, *Traité du burin*, 109.

12. Flocon, "Main ouvrière et main rêveuse," 63.

13. Flocon, *Traité du burin*, 109.

14. "Le philosophe et le graveur: Entretien entre Gaston Bachelard, Albert Flocon et Jean Amrouche," 11.

15. Hedwig Wingler, "Die Hand ist ein Auge: Zu den Arbeiten von Albert Flocon," in *Albert Flocon: Une poétique de la vision*, 33–42, here 42.

16. Bachelard, "Albert Flocon's 'Engraver's Treatise,'" 79.

17. Bachelard, "Albert Flocon's 'Engraver's Treatise,'" 82.

18. Flocon, *Traité du burin*, 103.

19. Flocon, "Éloge du burin," 13.

20. Bachelard, *Earth and Reveries of Will*, 27.

21. André Leroi-Gourhan, *L'homme et la matière* (Paris: Albin Michel, 1943), 19.

22. Bachelard, "Albert Flocon's 'Engraver's Treatise,'" 80, there translated as "implement-awareness."

23. Flocon, "Le philosophe et le graveur," 277.

24. Flocon, *Traité du burin*, 18.

25. Flocon, *Traité du burin*, 24.

26. Flocon, *Traité du burin*, 18.

27. Flocon, *Traité du burin*, 103.

POETICS

1. Bachelard, *Water and Dreams*, 1. On "material imagination," see also Natalie Adamson, "Black Flowers Blossom: Bachelard, Soulages and the Material Imaginary of Abstract Painting," in *Material Imagination: Art in Europe, 1946–72*, ed. Natalie Adamson and Steven Harris, 23–43 (Sussex: Wiley Blackwell, 2017).

2. Joseph Noiret, "Gaston Bachelard et Henri Lefebvre dans Cobra," in *Cobra en Fange: Vandercam-Dotrement, Dessin, écriture, matière (1958–1960)*, ed. Michel Draguet, 37–55 (Brussels: Groupe de Recherche en Art moderne, 1994), 45–47.

3. Bachelard, *Water and Dreams*, 3.

4. Bachelard, *Water and Dreams*, 2.

5. Bachelard, *Water and Dreams*, 3.

6. Bachelard, *Earth and Reveries of Will*, 1–2.

7. Bachelard, *Earth and Reveries of Will*, 6, there translated as "images of labor."

8. Bachelard, *Earth and Reveries of Will*, 1.

9. Bachelard, *Earth and Reveries of Will*, 29 (translation emended).

10. Bachelard, *Earth and Reveries of Will*, 1 and 4.

11. Bachelard, *Air and Dreams*, 11.

12. Bachelard, *Air and Dreams*, 247 and 248.

13. Bachelard, *Air and Dreams*, 250.

14. Bachelard, *Air and Dreams*, 149.

15. Gaston Bachelard, *The Poetics of Reverie: Childhood, Language, and the Cosmos*, trans. Daniel Russell (Boston: Beacon Press, 1969) (*La poétique de la rêverie*, 1960), 24.

16. Bachelard, *Poetics of Reverie*, 17.

17. Bachelard, *Poetics of Reverie*, 15. See also Michel Fichant, "Gaston Bachelard ou la philosophie et ses doubles," *Les Cahiers de Fontenay* 1 (1975): 100–14; and recently Véronique Le Ru, "Gaston Bachelard, philosophe du jour, philosophe de la nuit," *Les Cahiers du CIRLEP*, April 4, 2016, http://cirlep.hypotheses.org/1258.

18. Bachelard, *Psychoanalysis of Fire*, 2.

19. André Parinaud, "'La science est l'esthétique de l'intelligence' nous déclare le professeur Bachelard," *Arts* 347 (February 22, 1952): 1 and 7, here 7.

20. Jean-Jacques Wunenburger, "Une alternance éthique," in Wunenburger, *Gaston Bachelard*, 563–70, here 567.

21. François Dagognet, "M. Gaston Bachelard, philosophe de l'imagination," *Revue internationale de philosophie* 14 (1960): 32–42, here 34; see also François Dagognet, *Gaston Bachelard: Sa vie, son oeuvre, avec un exposé à sa philosophie* (Paris: Presses Universitaires de France, 1965).

22. Bachelard, *Earth and Reveries of Will*, 57.

23. Jean-Luc Pouliquen, "Un petit traité d'émerveillement," preface to Gaston Bachelard, *Lettres à Louis Guillaume* (Rennes: La Part Commune, 2009), 16.

24. Georges Canguilhem, "Sur une épistémologie concordataire," in Georges Bouligand et al., *Hommage à Gaston Bachelard: Études de philosophie et d'histoire des sciences* (Paris: Presses Universitaires de France, 1957), 3–12, here 10.

25. Canguilhem, "Sur une épistémologie concordataire," 3.

26. Roland Kuhn, *Über Maskendeutungen im Rorschachschen Versuch* (Basle: S. Karger, 1944); in French: *Phénoménologie du masque à travers le test de Rorschach*, trans. Jacqueline Verdeaux (Bruges: Desclée de Brouwer, 1957). Bachelard's preface to the translation, "Le masque," is translated as "The Mask" in Bachelard, *Right to Dream*, 176–87.

27. CAN 1080, Fonds Georges Canguilhem, CAPHÉS Centre d'Archives en Philosophie, Histoire et Édition des Sciences, Paris. The dedication is reproduced at www.bib.ens.fr/Encres.506.0.html.

28. Bachelard, "The Mask," 187.

29. Flocon, *Points de fuite, 2*, 189; see also Flocon, "Main ouvrière et main rêveuse," 59.

30. Albert Flocon, "Création—recréation," in *Art et science: de la créativité*, ed. Centre Culturel de Cerisy-la-Salle (Paris: Union Générale d'Éditions, 1972), 48–58, here 55.

CASTLES IN SPAIN

1. Flocon, *Points de fuite, 2*, 197.

2. Gaston Bachelard, introduction to Gaston Bachelard and Albert Flocon, *Châteaux en Espagne* (Paris: Cercle Grolier Les amis du livre moderne, 1957). Bachelard's text is translated as "Castles in Spain" in Bachelard, *Right to Dream*, 83–102, here 83.

3. Bachelard to Flocon, September 19, 1956 (Catherine Ballestero's archives).

4. This may explain why Flocon named 1956 as the year of printing. Judging by the dates given on the engravings (six of them are signed, four in 1952, one in 1953, and one in 1956), it seems that the plates were produced between 1952 and 1956, in other words a period not of two years, as Flocon reports, but four. See Flocon, *Points de fuite, 2*, 199.

5. Bachelard, "Castles in Spain," 83.

6. Bachelard, "Castles in Spain," 83.

7. Flocon, "Regard en arrière," 81.

8. Bachelard, "Castles in Spain," 85.

9. Bachelard, "Castles in Spain," 87 (translation emended).

10. Bachelard, *New Scientific Spirit*, 11–12.

11. Bachelard, *New Scientific Spirit*, 13.

12. See, for example, Bouligand et al., *Hommage à Gaston Bachelard*, frontispiece. Several sketches for the portrait are also reproduced in *Bulletin de l'Association des Amis d'Albert Flocon* (1999): 12–13, 17.

13. Cristina Chimisso, *Gaston Bachelard: Critic of Science and the Imagination* (London: Routledge, 2001).

14. Flocon, *Points de fuite, 2*, 198.

15. Bachelard, *Earth and Reveries of Will*, 147 (translation emended).

16. Bachelard, "Castles in Spain," 91.

17. Flocon, "Main ouvrière et main rêveuse," 63.

18. Flocon, "Le philosophe et le graveur," 271.

19. Gaston Bachelard, *Le rationalisme appliqué* (Paris: Presses Universitaires de France, 1949), 1 and 9.

20. Bachelard, *Le rationalisme appliqué*, 10. Sandra Pravica has proposed the term "the tentative" to reflect this issue. Sandra Pravica, *Bachelards tentative Wissenschaftsphilosophie* (Vienna: Passagen, 2015).

21. Bachelard, *Le rationalisme appliqué*, 43.

22. Bachelard, *Le rationalisme appliqué*, 11.

23. Bachelard, *Le rationalisme appliqué*, 49.

24. Bachelard, *Le rationalisme appliqué*, 214.

25. Albert Flocon, cited in Fulacher, "Albert Flocon," 34.

26. Bachelard, "Castles in Spain," 90–91.

27. Gaston Bachelard, "Noumenon and Microphysics" (1931–1932), trans. Bernard Roy, *Philosophical Forum* 37, no. 1 (2006): 75–84, here 79 (translation emended) and 84.

28. Bachelard, "Castles in Spain," 94–95.

29. Flocon, *Points de fuite, 2*, 198.

30. Flocon, "Main ouvrière et main rêveuse," 62.

31. Bachelard, "Castles in Spain," 83.

32. Bachelard, *Le rationalisme appliqué*, 54.

33. Flocon, *Points de fuite, 2*, 433.

34. Bachelard, "Castles in Spain," 101.
35. Bachelard, "Castles in Spain," 102.
36. Flocon, *Points de fuite, 2*, 198.

SCIENCE, ART, LITERATURE

1. Gaston Bachelard, "Introduction à la Bible de Chagall," in Marc Chagall, "Dessins pour la Bible," *Verve* 37–38 (1960): xiii. The text is translated as "Introduction to Chagall's Bible" in *Right to Dream*, 8–23. See also Barbara Puthomme, *Le rien profond: Pour une lecture bachelardienne de l'art contemporain* (Paris: L'Harmattan, 2003); Anthony Spiegeler, "Gaston Bachelard et les artistes: Une légitimité à double sens?" *Annales d'histoire de l'art et d'archéologie* 38 (2016): 161–68.

2. José Corti, *Rêves d'encre: Vingt-cinq images présentées par Paul Éluard, René Char, Julien Gracq et Gaston Bachelard* (Paris: Librairie José Corti, 1945). Second edition, with three additional drawings, Paris: Librairie José Corti, 1969. Bachelard's text, "Une rêverie de la matière," is translated as "A Dream of Matter" in Bachelard, *Right to Dream*, 49–51, here 49.

3. Marcoussis, *Les devins*.

4. Gaston Bachelard, "Marcoussis et ses poètes," in *Hommage à Marcoussis: Mai 1962, Galerie Kriegel* (Paris: Galerie Kriegel, 1962), 9–10.

5. Tristan Tzara, *Indicateur des chemins de coeur: Eaux-fortes de Louis Marcoussis* (Paris: Éditions Jeanne Bucher, 1928); Louis Marcoussis, *Planches de salut: 10 gravures à l'eau forte et au burin*, with a preface by Tristan Tzara (Paris: Éditions Jeanne Bucher, 1931).

6. *Christine Boumeester*, with texts by Francis Picabia, Gaston Bachelard, Max Clarac-Serou, Noël Arnaud, and Iaroslav Serpan (Paris: Éditions Instance, 1951).

7. Bachelard to Flocon, Paris, October 23, 1951 (Catherine Ballestero's archives).

8. *Christine Boumeester, peintures*, Galerie Kerchache, art moderne, 3, rue des Beaux-Arts, Paris 6, no date (exhibition dates March 30 to April 30, 1962).

9. *Jacques Fouquet, Blancs et noirs—peintures.* Galérie Montjoie, 10, rue Jean-du-Bellay, Paris 4, with a text by Gaston Bachelard (exhibition dates May 21 to June 11, 1953).

10. Bachelard, "Le cosmos du fer," in "Chillida," *Derrière le miroir* 90–91 (1956). The text is translated as "The Cosmos of Iron" in *Right to Dream*, 44–48.

11. *Un document sur "Le long voyage" et les autres tapisseries d'Asger Jorn & Pierre Wemaëre. Avec deux présentations de Gaston Bachelard et Michèle Bernstein*, Galerie Les Quatre Saisons (Paris: Éditions La Bibliothèque d'Alexandrie, 1960).

12. Chagall, "Dessins pour la Bible." As early as 1952, for the appearance of an edition of La Fontaine's fables illustrated by Chagall (Paris: Tériade Éditeurs, 1951/1952), Maeght had published a booklet of Chagall's lithographs with a text by Bachelard, "La lumière des origines," in "Chagall," *Derrière le miroir* 44–45 (1952). Bachelard's text is translated under the erroneous title "The Origins of Light" in *Right to Dream*, 24–28.

13. Jean Rostand, *Notes d'un biologiste, gravures au burin par Albert Flocon* (Paris: Les pharmaciens bibliophiles, 1954). See also the earlier publication of a series of "notes," Jean Rostand, "Notes d'un biologiste," *La nouvelle revue française* 27, no. 314 (1939): 693–98. The same issue of this journal featured an essay by Bachelard, "Le bestiaire de Lautréamont," 711–34.

14. Bachelard to Flocon, September 8, 1950 (Catherine Ballestero's archives).

15. Flocon, *Points de fuite, 2*, 195–57.

16. Rostand, *Notes d'un biologiste*, 85.

17. Louis de Broglie, *Que sommes-nous, où allons-nous? Maquette et gravures sur bois par Albert Flocon* (Paris: Éditions Estienne, 1956). See also Flocon, *Points de fuite, 2*, 215–17.

18. Albert Flocon, *L'univers des livres: Étude historique des origines à la fin du XVIIIe siècle* (Paris: Hermann, 1961).

19. Flocon, *Points de fuite, 2*, 189.

20. Albert Flocon and René Taton, *La perspective* (Paris: Presses Universitaires de France, 1963).

21. Flocon, *Points de fuite, 2*, 308.

22. André Barre and Albert Flocon, *La perspective curviligne: De l'espace visuel à l'image construite* (Paris: Flammarion, 1968). Published in English as Albert Flocon and André Barre, *Curvilinear Perspective*, trans. Robert Hansen (Berkeley: University of California Press, 1987).

23. Wingler, "Die Hand ist ein Auge," 34.

24. Yves Michaud, "Le labyrinthe des possibles," in *Albert Flocon, Perspectives. Exposition du 23 septembre au 6 novembre 1994, École nationale supérieure des Beaux-Arts* (Paris: École nationale supérieure des Beaux-Arts, 1994), 2–3, here 2.

25. Flocon, "Main ouvrière et main rêveuse," 64.

26. Bachelard, *Earth and Reveries of Will*, 4.

27. Bachelard, *Earth and Reveries of Will*, 5 (translation emended).

28. Bachelard, *Poetics of Reverie*, 17.

29. Bachelard, *Earth and Reveries of Will*, 5.

30. Bachelard, *Water and Dreams*, 10 (translation emended).

31. Bachelard, *Le rationalisme appliqué*, 56.

BIBLIOGRAPHY

Adamson, Natalie. "Black Flowers Blossom: Bachelard, Soulages and the Material Imaginary of Abstract Painting." In *Material Imagination: Art in Europe, 1946–72*, edited by Natalie Adamson and Steven Harris, 23–43. Sussex: Wiley Blackwell, 2017.

Bachelard, Gaston. *Essai sur la connaissance approchée.* Paris: Vrin, 1928.

Bachelard, Gaston. *Étude sur l'évolution d'un problème de physique: La propagation thermique dans les solides.* Paris: Vrin, 1928.

Bachelard, Gaston. "Noumenon and Microphysics" (1931–1932). Translated by Bernard Roy. *Philosophical Forum* 37, no. 1 (2006): 75–84.

Bachelard, Gaston. *The New Scientific Spirit.* Translated by Arthur Goldhammer. Boston: Beacon Press, 1984 (*Le nouvel esprit scientifique*, 1934).

Bachelard, Gaston. *The Formation of the Scientific Mind: A Contribution to a Psychoanalysis of Objective Knowledge.* Translated by Mary McAllester Jones. Manchester: Clinamen, 2002 (*La formation de l'esprit scientifique*, 1938).

Bachelard, Gaston. *The Psychoanalysis of Fire.* Translated by Alan C. M. Ross. Boston: Beacon Press, 1964 (*La psychanalyse du feu*, 1938).

Bachelard, Gaston. *The Philosophy of No: A Philosophy of the New Scientific Mind.* Translated by G. C. Waterston. New York: Orion Press, 1968 (*La philosophie du non*, 1940).

Bachelard, Gaston. *Water and Dreams: An Essay on the Imagination of Matter.* Translated by Edith R. Farrell. Dallas: The Dallas Institute of Humanities and Culture, 1983 (*L'eau et les rêves: Essai sur l'imagination de la matière*, 1942).

Bachelard, Gaston. *Air and Dreams: An Essay on the Imagination of Movement*. Translated by Edith R. Farrell and C. Frederick Farrell. Dallas: The Dallas Institute of Humanities and Culture, 1988 (*L'air et les songes: Essai sur l'imagination du mouvement*, 1943).

Bachelard, Gaston. *Earth and Reveries of Repose: An Essay on Images of Interiority*. Translated by Mary McAllester Jones. Dallas: The Dallas Institute of Humanities and Culture, 2011 (*La terre et les rêveries du repos: Essai sur les images de l'intimité*, 1948).

Bachelard, Gaston. *Earth and Reveries of Will: An Essay on the Imagination of Matter*. Translated by Kenneth Haltman. Dallas: The Dallas Institute of Humanities and Culture, 2002 (*La terre et les rêveries de la volonté: Essai sur l'imagination des forces*, 1948).

Bachelard, Gaston. *Le rationalisme appliqué*. Paris: Presses Universitaires de France, 1949.

Bachelard, Gaston. "La lumière des origines." In "Chagall," *Derrière le miroir* 44–45 (1952): no pagination.

Bachelard, Gaston. "Le cosmos du fer." In "Chillida," *Derrière le miroir* 90–91 (1956): no pagination.

Bachelard, Gaston. "Introduction à la Bible de Chagall." In Marc Chagall, "Dessins pour la Bible," *Verve* 37–38 (1960): xiii.

Bachelard, Gaston. *The Poetics of Reverie: Childhood, Language, and the Cosmos*. Translated by Daniel Russell. Boston: Beacon Press, 1969 (*La poétique de la rêverie*, 1960).

Bachelard, Gaston. *The Flame of a Candle*. Translated by Joni Caldwell. Dallas: The Dallas Institute of Humanities and Culture, 1988 (*La flamme d'une chandelle*, 1961).

Bachelard, Gaston. "Marcoussis et ses poètes." In *Hommage à Marcoussis: Mai 1962, Galerie Kriegel*, 9–10. Paris: Galerie Kriegel, 1962.

Bachelard, Gaston. *The Right to Dream*. Translated by J. A. Underwood. New York: Orion, 1971 (*Le droit de rêver*, 1970).

Bachelard, Gaston. *Fragments of a Poetics of Fire*. Translated by Kenneth Haltman. Dallas: The Dallas Institute of Humanities and Culture, 1990 (*Fragments d'une poétique du feu*, 1988).

Bachelard, Gaston, and Albert Flocon. *Paysages*. Rolle: Eynard, 1950.

Bachelard, Gaston, and Albert Flocon. *Châteaux en Espagne*. Paris: Cercle Grolier, Les amis du livre moderne, 1957.

Bachelard, Gaston, Albert Flocon, and Jean Amrouche. "Le philosophe et le graveur: Entretien entre Gaston Bachelard, Albert Flocon et Jean Amrouche (mai 1950, mai 1953)." *Bulletin de l'Association des Amis d'Albert Flocon* (1999): 4–18.

Bachelard, Gaston, Paul Éluard, Jean Lescure, Henri Mondor, Francis Ponge, René de Solier, Tristan Tzara, and Paul Valéry. Engravings by Christine Boumeester, Roger Chastel, Pierre Courtin, Sylvain Durand, Jean Fautrier, Marcel Fiorini, Albert Flocon, Henri Goetz, Léon Prébandier, Germaine Richier, Jean Signovert, Raoul Ubac, Roger Vieillard, Jacques Villon, Gérard Vulliamy, and Albert-Edgar Yersin. *À la gloire de la main*. Paris: Aux dépens d'un amateur, 1949.

Barre, André, and Albert Flocon. *La perspective curviligne: De l'espace visuel à l'image construite*. Paris: Flammarion, 1968.

Béhar, Henri, ed. *Du Surréalisme au Front Populaire: Inquisitions*. Paris: Éditions du CNRS, 1990.

Boehm, Gottfried. "Bildsinn und Sinnesorgane." *Neue Hefte für Philosophie* 18/19 (1980): 118–32.

de Broglie, Louis. *Que sommes-nous, où allons-nous? Maquette et gravures sur bois par Albert Flocon*. Paris: Éditions Estienne, 1956.

Canguilhem, Georges. "Sur une épistémologie concordataire." In Georges Bouligand et al., *Hommage à Gaston Bachelard: Études de philosophie et d'histoire des sciences*, 3–12. Paris: Presses Universitaires de France, 1957.

Chimisso, Cristina. *Gaston Bachelard: Critic of Science and the Imagination*. London: Routledge, 2001.

Chiore, Valeria. "Force, provocation, volonté: *Paysages. Notes d'un philosophe pour un graveur*, entre ontologie des éléments et phénoménologie de la parole poétique." *Ideação* 25, no. 1 (2011): 121–36.

Chiore, Valeria. "Bachelard-Flocon, *Paysages*: Il mouvement—Forza, Provocazione, Volontà tra *éléments* e *poésie*." In "Bachelard e la plasticità della materia." Special issue, *Altre Modernità* (2012): 76–88.

Christine Boumeester. With texts by Francis Picabia, Gaston Bachelard, Max Clarac-Serou, Noël Arnaud, and Iaroslav Serpan. Paris: Éditions Instance, 1951.

Christine Boumeester, Peintures. Paris: Galerie Kerchache, n.d. [1962]. Exhibition catalog.

Corti, José. *Rêves d'encre: Vingt-cinq images présentées par Paul Éluard, René Char, Julien Gracq et Gaston Bachelard.* Paris: Librairie José Corti, 1945.

Dagognet, François. "M. Gaston Bachelard, philosophe de l'imagination." *Revue internationale de philosophie* 14 (1960): 32–42.

Dagognet, François. *Gaston Bachelard: Sa vie, son oeuvre, avec un exposé à sa philosophie.* Paris: Presses Universitaires de France, 1965.

Didi-Huberman, Georges. *La ressemblance par contact: Archéologie, anachronisme et modernité de l'empreinte.* Paris: Minuit, 2008.

École des Beaux-Arts de Metz, ed. *Albert Flocon: Une poétique de la vision. Du Bauhaus à la perspective curviligne / Albert Flocon: Eine Poetik des Sehens. Vom Bauhaus zur kurvilinearen Perspektive.* Metz: École des Beaux-Arts, 1992.

Éluard, Paul. *Le livre ouvert II.* Paris: Éditions Cahiers d'Art, 1942.

Éluard, Paul, Gaston Bachelard, and Albert Flocon. *Die Bücher des Albert Flocon.* Translated by Nicolaus Bornhorn. Düsseldorf: Bollmann, 1991.

Éluard, Paul, and Albert Flocon. *Perspectives: Poèmes sur des gravures de Albert Flocon.* Paris: Maeght, 1948.

Fichant, Michel. "Gaston Bachelard ou la philosophie et ses doubles." *Cahiers de Fontenay* 1 (1975): 100–14.

Filloux, Jean-Claude. "Témoignage sur la vie de Gaston Bachelard." In *Gaston Bachelard: Science et poétique, une nouvelle éthique?*, edited by Jean-Jacques Wunenburger, 493–502. Paris: Hermann, 2013.

Flocon, Albert. "L'éloge du burin." *Art d'aujourd'hui* 9 (April 1950): 13.

Flocon, Albert. *Traité du burin. Illustré par l'auteur. Préface de Gaston Bachelard.* Paris: Librairie Auguste Blaizot, 1952.

Flocon, Albert. *L'univers des livres: Étude historique des origines à la fin du XVIIIe siècle.* Paris: Hermann, 1961.

Flocon, Albert. "Création—recréation." In *Art et science: de la créativité*, edited by Centre Culturel de Cerisy-la-Salle, 48–58. Paris: Union Générale d'Éditions, 1972.

Flocon, Albert. "Le philosophe et le graveur." In *Bachelard, Colloque de Cerisy*, 271–78. Paris: Union Générale d'Éditions, 1974.

Flocon, Albert. "Main ouvrière et main rêveuse (Entretien avec Gil Jouanard)." In "Bachelard ou le droit de rêver." Special issue, *Solaire* 10 (1983): 57–65.

Flocon, Albert. *Suites expérimentales, 1943–1983.* Vienna: Medusa, 1983.

Flocon, Albert. *Scénographies au Bauhaus, Dessau, 1927–1930: Hommage à Oskar Schlemmer en plusieurs tableaux.* Paris: Archimbaud, 1987.

Flocon, Albert. "Verflechtungen." In *Prägungen: Deutsche in Paris*, edited by Georg Lechner, 284–98. Düsseldorf: Bollmann, 1991.

Flocon, Albert. "Regard en arrière." In *Albert Flocon: Une poétique de la vision. Du Bauhaus à la perspective curviligne / Albert Flocon: Eine Poetik des Sehens. Vom Bauhaus zur kurvilinearen Perspektive*, edited by École des Beaux-Arts de Metz, 77–83. Metz: École des Beaux-Arts, 1992.

Flocon, Albert. *Points de fuite, 1: 1909–1933.* Neuchâtel: Éditions Ides et Calendes, 1994.

Flocon, Albert. *Points de fuite, 2: 1933–1994.* Neuchâtel: Éditions Ides et Calendes, 1995.

Flocon, Albert, and René Taton. *La perspective.* Paris: Presses Universitaires de France, 1963.

Focillon, Henri. "In Praise of Hands" (1939). Translated by S. L. Faison. In Henri Focillon, *The Life of Forms in Art*, translated by Charles Beecher Hogan, George Kubler, and S. L. Faison, 157–85. New York: Zone Books, 1989.

Fulacher, Pascal. "Albert Flocon: Graveur et bibliomane." *Art et métiers du livre* 135 (June 1985): 33–36.

Gateau, Jean-Charles. *Éluard, Picasso et la peinture, 1936–1952.* Geneva: Librarie Droz, 1983.

Gasché, Rodolphe. "Roundtable on Translation." In *The Ear of the Other: Otobiography, Transference, and Translation. Texts and Discussions with Jacques Derrida*, edited by Christie McDonald, translated by Peggy Kamuf and Avital Ronell, 93–161. Lincoln: University of Nebraska Press, 1988.

Herzogenrath, Wulf. "Von der Bauhaus-Bühne zum kurvenlinearen Raumbild: Albert Flocon." In *Das Bauhaus und Frankreich: Le Bauhaus et la France. 1919–1940*, edited by Isabelle Ewig, Thomas W. Gaehtgens, and Matthias Noell, 465–78. Berlin: Akademie, 2002.

Jacques Fouquet, Blancs et noirs—peintures. With a text by Gaston Bachelard. Paris: Galérie Montjoie, n.d. [1953]. Exhibition catalog.

Kandinsky, Wassily. *Point and Line to Plane* (1926). Translated by Howard Dearstyne and Hilla Rebay. New York: Solomon R. Guggenheim Foundation, 1947.

Kuhn, Roland. *Über Maskendeutungen im Rorschachschen Versuch.* Basle: S. Karger, 1944.

Kuhn, Roland. *Phénoménologie du masque à travers le test de Rorschach.* Translated by Jacqueline Verdeaux. Bruges: Desclée de Brouwer, 1957.

Leroi-Gourhan, André. *L'homme et la matière.* Paris: Albin Michel, 1943.

Le Ru, Véronique. "Gaston Bachelard, philosophe du jour, philosophe de la nuit." *Les Cahiers du CIRLEP*, April 4, 2016, http://cirlep.hypotheses.org/1258.

Marcoussis, Louis. *Planches de salut: 10 gravures à l'eau forte et au burin.* Preface by Tristan Tzara. Paris: Éditions Jeanne Bucher, 1931.

Marcoussis, Louis. *Les devins: 16 Pointes sèches.* Text by Gaston Bachelard, preface by Maurice Raynal. Paris: La Hune, 1946.

Margolin, Jean-Claude. *Bachelard*. Paris: Seuil, 1974.

von Maur, Karin. "Oskar Schlemmer as Choreographer and Stage Designer." In *Oskar Schlemmer: Visions of a New World*, edited by Ina Conzen and Staatsgalerie Stuttgart, 191–246. Munich: Hirmer, 2014.

Mentzel, Albert. *Villages de France: Jeu de construction*. Paris: Flammarion, 1937.

Mentzel, Albert. *Le cirque animé*. Paris: Flammarion, 1938.

Mentzel, Albert, and Albert Roux, eds. *Formes nues*. Paris: Éditions d'Art Graphique et Photographique, 1935.

Michaud, Yves. "Le labyrinthe des possibles." In *Albert Flocon, Perspectives. Exposition du 23 septembre au 6 novembre 1994, École nationale supérieure des Beaux-Arts*, 2–3. Paris: École nationale supérieure des Beaux-Arts, 1994.

Nasim, Omar W. *Observing by Hand: Sketching the Nebulae in the Nineteenth Century*. Chicago: University of Chicago Press, 2013.

Noiret, Joseph. "Gaston Bachelard et Henri Lefebvre dans Cobra." In *Cobra en Fange: Vandercam-Dotrement, Dessin, écriture, matière (1958–1960)*, edited by Michel Draguet, 37–55. Brussels: Groupe de Recherche en Art moderne, 1994.

Parinaud, André. *Gaston Bachelard*. Paris: Flammarion, 1996.

Puthomme, Barbara. *Le rien profond: Pour une lecture bachelardienne de l'art contemporain*. Paris: L'Harmattan, 2003.

Ray, Man, and Paul Éluard. *Les mains libres: Dessins de Man Ray illustrés par les poèmes de Paul Éluard*. Paris: Éditions Jeanne Bucher, 1937.

Rostand, Jean. "Notes d'un biologiste." *La nouvelle revue française* 27, no. 314 (1939): 693–98.

Rostand, Jean. *Notes d'un biologiste, gravures au burin par Albert Flocon*. Paris: Les pharmaciens bibliophiles, 1954.

de Rougemont, Denis. *Penser avec les mains*. Paris: Albin Michel, 1936.

Spiegeler, Anthony. "Gaston Bachelard et les artistes: Une légitimité à double sens?" *Annales d'histoire de l'art et d'archéologie* 38 (2016): 161–68.

Starobinski, Jean. "Paysages gravés." *Journal de Genève* (January 27, 1951): 3–4.

Tzara, Tristan. *Indicateur des chemins de coeur: Eaux-fortes de Louis Marcoussis*. Paris: Éditions Jeanne Bucher, 1928.

Tzara, Tristan. *Grains et issues* (1935). Edited with an introduction and notes by Henri Béhar. Paris: Garnier-Flammarion, 1981.

Un document sur "Le long voyage" et les autres tapisseries d'Asger Jorn & Pierre Wemaëre. With texts by Gaston Bachelard and Michèle Bernstein. Paris: Galérie Les Quatre Saisons / Éditions La Bibliothèque d'Alexandrie, 1960.

Vernay, Anne, and Richard Walter, eds. *La main à plume: Anthologie du surréalisme sous l'occupation*. Preface by Gérard Durozoi. Paris: Syllepse, 2008.

Warnke, Martin. "Der Kopf in der Hand." In Martin Warnke, *Nah und Fern zum Bilde: Beiträge zu Kunst und Kunsttheorie*, edited by Michael Diers, 108–20. Cologne: Dumont, 1997.

Wingler, Hedwig. "Die Hand ist ein Auge: Zu den Arbeiten von Albert Flocon." In *Albert Flocon: Une poétique de la vision. Du Bauhaus à la perspective curviligne / Albert Flocon: Eine Poetik des Sehens. Vom Bauhaus zur kurvilinearen Perspektive*, edited by École des Beaux-Arts de Metz, 33–42. Metz: École des Beaux-Arts, 1992.

Wunenburger, Jean-Jacques. "Une alternance éthique." In *Gaston Bachelard: Science et poétique, une nouvelle éthique?*, edited by Jean-Jacques Wunenburger, 563–70. Paris: Hermann, 2013.

Zwijnenberg, Robert. "Rafaël zonder handen: Enige voorbereidende opmerkingen over een esthetica met handen." *Kunstlicht* 17, no. 1 (1996): 23–28.

INDEX

Art d'aujourd'hui, 43
Auschwitz, 6
Air and Dreams, 52
Albers, Josef, 5, 13
Aragon, Louis, 1
Atelier de l'Ermitage, 22
Atelier Georges Visat, 9

Bachelard, Gaston
  art, 77–78
  biography, 6–8
  Castles in Spain, 57–76
  Engraving, On, 43–50
  epistemology, 43, 54, 60, 68
  Landscapes, 23–41
  literature, 79–80
  poetics, 51–56
  Praise of hands, In, 17–22
  surrationalism, 1–4
  surrealism, 2, 8, 12, 79
Bachelard, Suzanne, 7, 33
Banzet, Paul, 57
Bar-sur-Aube, 6–7
Barre, André, 79
Bauhaus, 5, 13, 20, 70
Berlin, 5–6
Blaizot, Auguste, 44
Blaizot, Georges, 10, 44
Boehm, Gottfried, 39
Bois-le-Roi, 69
Bollmann, Stefan, 21
Boumeester, Christine, 20, 77
Breton, André, 12, 18
Broglie, Louis de, 78
Brunschvicg, Léon, 2, 7
Bucher, Jeanne, 12
Buffon, Georges, 49

Caillois, Roger, 1–2
Canguilhem, Georges, 55–56
Cerisy-la-Salle, 49
Chagall, Marc, 78
Châteaux en Espagne (Castles in Spain), 57, 59, 61, 64, 71, 75–76
Chillida, Eduardo, 78
Chimisso, Cristina, 64
Cobra, 23, 51
Corti, José, 15, 17, 77

Dagognet, François, 54
Davy, Georges, 7
Dessau, 5–6
Derrière le miroir, 9, 78
Dessins pour la bible, 78
Devins, Les, 77

Didi-Huberman, Georges, 22
Dijon, 7–8, 23, 31, 41, 59
Dijon, University of, 7
Döbeln, 5
Drancy, 6
Duhamel du Monceau, Henri-Louis, 40

Earth and Reveries of Repose, 17, 40
Earth and Reveries of Will, 15, 17, 49, 52, 64, 79–80
École Estienne, 78–79
Éluard, Paul, 3, 10–13, 16, 20, 49
Escher, Maurits Cornelis, 78
Eynard, Paul, 25

Faucher, Paul, 6
Fequet and Baudier (printers), 25, 57
Filloux, Marius, 7
Flame of a Candle, The, 15
Flammarion, 7
Flocon, Albert
  biography, 5–6
  Castles in Spain, 57–76
  Engraving, On, 43–50
  Landscapes, 23–41
  memoirs, 5, 13, 17, 21, 24, 56, 69; see also *Points de fuite*
  Perspectives, 9–14
  Praise of hands, In, 17–22
  science, 78–79
Flocon, Ferdinand, 6
Focillon, Henri, 11–13, 49, 80
Formation of the Scientific Mind, The, 7
Fouquet, Jacques, 78
Frankfurt, 6
Frazier-Soye, 44
Friedländer, Johnny, 22
Fulacher, Pascal, 24

Galerie des Deux, 17
Galerie Kriegel, 77
Gasché, Rodolphe, 39
Gloire de la main, À la (In praise of hands), 18, 20–21, 43
Goetz, Henri, 9, 13, 20
Grains et issues, 1
Grolier, Cercle, 57
Grolier, Jean, 57
Guillaume, Louis, 54

Haubinda, 5
Hommage à Gaston Bachelard, 55
Homme et la matière, 49
Hugo, Victor, 49

Imprimerie Union, 9

Jorn, Asger, 23, 78
Jouanard, Gil, 66

Kandinsky, Wassily, 5
Köpenick, 5
Kuhn, Roland, 55

Le Bon, Gustave, 7
Leblanc, Georges, 18, 21–22, 25, 44, 57
Lequenne, Fernand, 40
Leroi-Gourhan, André, 49

Maeght, Aimé, 9, 78
Main à plume, La, 13
Mains libres, Les, 12
Marcoussis, Louis, 77

Maskendeutung im Rorschachschen Versuch, Über (On the interpretation of masks in the Rorschach Test), 55
Mentzel, Albert (later Albert Flocon), 5–6
Metz, 47
Michaud, Yves, 79
Monnerot, Jules-Marcel, 1
Montrouge, 6

New Scientific Spirit, The, 2, 7, 60
Nice, 27
Nieuwenhuys, Constant, 23
Noiret, Joseph, 51
Notes d'un biologiste, 78
Nouvelle revue française, La, 2

Paris, 6–9, 12, 17, 25, 44, 77–78
Paysages (Landscapes), 23–25, 32, 26, 57, 61–62
Penser avec les mains, 12
Perspectives, 9–10, 13–14, 22, 25, 44
Philosophy of No, The, 8
Pibrac, 6
Poetics of Reverie, The, 8
Points de fuite, 70; see also Flocon, memoirs
Pouliquen, Jean-Luc, 54
Psychoanalysis of Fire, The, 3, 7, 15

Rationalisme appliqué, Le, 68
Ray, Man, 6, 12
Raynor, Dorka, 12
Rèves d'encre, 77
Rey, Abel, 7
Rimbaud, Arthur, 13
Rolle, 25
Rossi, Jeanne, 7
Rostand, Jean, 78
Rothschild, Lotte, 6
Rougemont, Denis de, 12

Scénographies au Bauhaus, 5
Scheler, Lucien, 10
Schlemmer, Oskar, 5, 70
Sorbonne, 7–8
Starobinski, Jean, 25

Table ronde, La, 12
Taton, René, 78
Toulouse, 6, 9
Traité du burin, Le (On engraving), 43–44, 47, 49
Tzara, Tristan, 1–2, 20, 49, 77

Ubac, Raoul, 20

Valéry, Paul, 21
Vallage, 7
Varga, Lucie, 6
Vasarely, Victor, 6
Verve, 78
Vie des formes, La, 11
Ville-d'Avray, 78
Viroflay, 6
Visat, Georges, 9; see also Atelier Georges Visat
Vulliamy, Gérard, 20

Wahl, Jean, 2
Water and Dreams, 51, 80
Wemaëre, Pierre, 78
Wingler, Hedwig, 47, 79
Wunenburger, Jean-Jacques, 54

Yersin, Albert-Edgar, 17, 25

www.ingramcontent.com/pod-product-compliance
Lightning Source LLC
LaVergne TN
LVHW040759070826
844660LV00025B/1194

* 9 7 8 1 4 3 8 4 7 2 1 0 2 *